ROUTLEDGE LIBRARY EDITIONS:
SOCIAL AND CULTURAL GEOGRAPHY

Volume 2

THE MAKERS OF
MODERN GEOGRAPHY

T0262784

THE MAKERS OF
MODERN GEOGRAPHY

ROBERT E. DICKINSON

Routledge
Taylor & Francis Group

LONDON AND NEW YORK

First published in 1969

This edition first published in 2014
by Routledge
2 Park Square, Milton Park, Abingdon, Oxfordshire OX14 4RN

and by Routledge
711 Third Avenue, New York, NY 10017

Routledge is an imprint of the Taylor and Francis Group, an informa business

First issued in paperback 2015

British Library Cataloguing in Publication Data
A catalogue record for this book is available from the British Library

ISBN 978-0-415-83447-6 (Set)
eISBN 978-1-315- 84860-0 (Set)
ISBN 978-0-415-73130-0 (hbk) (Volume 2)
ISBN 978-1-138-98954-2 (pbk) (Volume 2)
ISBN 978-1-315-84859-4 (ebk) (Volume 2)

Publisher's Note
The publisher has gone to great lengths to ensure the quality of this reprint but points out that some imperfections in the original copies may be apparent.

Disclaimer
The publisher has made every effort to trace copyright holders and would welcome correspondence from those they have been unable to trace.

The Makers of
Modern Geography

by

ROBERT E. DICKINSON

LONDON

Routledge & Kegan Paul

First published 1969
by Routledge & Kegan Paul Limited
Broadway House, 68—74 Carter Lane
London, E.C.4
Printed in Great Britain
by C. Tinling & Co. Ltd
Liverpool, London & Prescot
SBN 7100 6288 5

To
Mary

Contents

Contents

PART THREE: *France*

List of Plates

Between pages 146 and 147

Acknowledgements

Acknowledgements are due to the Association of American Geographers for permission to reproduce a quotation from *The Nature of Geography*, by R. Hartshorne, 1939, and to the following for permission to include the photographs:

'Berichte' (Partsch).
Bildarchiv d. Österreichische Nationalbibliothek (Von Humboldt).
British Museum (Frontispiece).
Cahiers de géographie de Québec (Blanchard).
Deutsche Rundschau für Geographie und Statistik (Ratzel).
Erdkunde Archiv für wissenschaftliche Geographie (Troll and Lautersach).
Faculté des Lettres de Strasbourg (Baulig).
Heidelberger Geographische Arbeiten (Hettner).
Institut de Géographie (Sorre).
Institut für Landeskunde (Gradmann, Hassinger, Penck, Philipson, Schlüter and Waibel).
Manchester University Press: Freeman *The Geographer's Craft* (Vidal de la Blache).
Royal Geographical Society (Ritter and von Richthofen).

Preface

The purpose of this book is to trace the development of modern geography as an organised body of knowledge in the light of the works of its foremost German and French contributors.

Geography as an explicitly defined field of knowledge is more than two thousand years old. The writings of Ptolemy and Strabo contain clear statements of basic geographical concepts, though we should always remember (as classical scholars so often forget) that their theoretical views, like those of their successors until our own day, suffered from a lack of factual information and maps about the lands, seas, and peoples of the earth that were the subject of their writings.

With the works of the early scholars, however, I am not directly concerned. Our purpose is to trace the development of geography over the past one hundred years or more. In this period it has experienced a quite prodigious development, as is evidenced by its full recognition in Universities throughout the civilised world, and by the activities of professional societies and international organisations.

Such being the purpose of the book, I may briefly outline its contents. The founders of modern geography were by international acclaim Alexander von Humboldt and Carl Ritter in the first half of the nineteenth century. There was a period thereafter that lasted for several decades (essentially the third quarter of the century) in which the geography of Ritter was treated in both Germany and France as an adjunct of history. The subject was of wide popularity in the field of exploration and was recognised as worthy of status in the University in the last quarter of the century—a development due both to the advancement of knowledge as well as to an active phase of colonial expansion in both countries.

The first chairs were established in the last quarter of the century and their holders were thus the first generation of University geographers. The leaders were Ferdinand von Richthofen and Friedrich Ratzel.

The second generation were students or followers of the first and their careers overlap with their teachers and their first pupils. This group includes Vidal de la Blache in France; Albrecht Penck, Alfred Hettner, Otto Schlüter and Joseph Partsch in Germany; Halford Mackinder in Britain; William Morris Davis in the United States; and Sten de Geer in Sweden. Their period of professional activity extends from the last two decades of the nineteenth century to the end of the first quarter of the twentieth century. All of them in the first part of their careers were contemporary with the elder masters, but remained at their posts for about twenty years after.

The third generation, which has many distinguished names, embraces the pupils of these men, or students who were greatly influenced by them, or, in a few cases, others who entered the professional field as what the Germans call *Autodiktaten*, that is, self-trained geographers. The period of this generation roughly fills the second quarter of this century and their most active years were between the wars. It is, indeed, convenient in many ways to regard the second and third generations as separated by the hiatus of the First World War.

The fourth generation includes the scholars who began their careers between the wars, have been leaders over the past twenty or thirty years, and are now reaching the end of their careers or are already in, or approaching, retirement.

A fifth generation includes those who since the Second World War have been actively engaged in research and now have twenty or thirty years of leadership before them. I have chosen not to select and deal with these scholars individually, but to examine the general trends of research and publication in the post-war years.

This chronological growth conditions the sequence of the chapters of the book, which, in its final draft, is limited to the German and French contributors, treated in two separate parts.

Several general comments are necessary. Geography as a University study is a hundred years old, but in both Britain and America there is widespread doubt as to what it is all about. My aim is to present to English-speaking scholars the substance of a long and respected heritage. If geography is what geographers do, let us look at the works of its great makers. This procedure must result in a certain amount of repetition, for which I offer no apologies. This fact certainly reveals the consistency of idea and purpose of its major makers. Further, I wish to state here at the outset that the consistent objective of all the outstanding makers of modern geography, European and American alike, to say nothing of Ptolemy and Strabo, is that their concern is with the regional concept, that is,

Preface

with the modes of association of terrestrial phenomena as distinctive specific and generic segments of the surface of the earth. I have quite deliberately excluded all reference to the development of ancillary fields of study, be it geomorphology or geopolitics, and selected the contributors and their works in the light of their relevance to the problems and procedures of the regional concept.

Professor Ph. Pinchemel, formerly at the University of Lille, and since 1965 at the Sorbonne, and Professor F. Wilhelm at the University of Kiel, kindly presented short reports on post-war trends. Professor Carl Troll provided invaluable data on the career of Albrecht Penck. Professors Jean Gottman and André Cholley have been very helpful on questions of geography in France. Dr. Richard Hartshorne made a careful reading of the third draft, and most of his comments on detail and relevance have been followed in a draft that has radically shifted in emphasis from the field of modern geography to a series of biographies. To all these I wish to convey my warmest thanks. Responsibility, however, for every word of the text rests with me, the author. Finally, much work has been put into the preparation of this book at its different stages by my wife and the final product is an expression of gratitude for her long enduring patience and sympathetic support.

ROBERT E. DICKINSON

Basic References

The basic references on the field of geography are as follows:

1. A. Hettner, *Die Geographie, Ihre Geschichte, Ihr Wesen und Ihre Methoden*, Breslau, 1927.
2. R. Hartshorne, *The Nature of Geography*, two volumes published in the *Annals of the Assoc. Amer. Geogrs.*, Vol. XXIX, September 1939 and December 1939, pp. 173-658. Reprinted as a single volume by the Association in 1946. Also *Perspective on the Nature of Geography*, Chicago and London, 1959. These two works are the most scholarly analysis of the field of geography in English, and probably in any language. They are based primarily on a thorough and critical survey of the European, and especially, the German contribution.
3. G. de Jong, *Chorological Differentiation as the Fundamental Principle of Geography*, Groningen, 1962.
4. G. Taylor, *Geography in the Twentieth Century*, London, 1951, is a collection of essays by some dozen authors from various countries on 'the evolution of geography', 'environment as a factor', and 'special fields'.
5. T. W. Freeman, *A Hundred Years of Geography*, London, 1961. Mainly a record of development in Britain, with very sporadic references to Continental and American contributions.

Part One

THE FOUNDERS

1

From Strabo to Kant

PTOLEMY AND STRABO

Descriptive writings of land and people are found in the oral traditions of classical Greece and are reflected in the works of Homer. A major step forward, in the realm of science, was taken by Eratosthenes at Alexandria in the third century before Christ. He not only carried out a remarkably accurate and well-known calculation of the dimensions of the earth but he also wrote a descriptive work called the *Geographica*. Although the work is no longer extant, it is known to have contained the first recorded use of the word 'geography'. The word is derived from the Greek *ge*, meaning the earth, and *grapho*, meaning I write. Geography literally means, therefore, writing about, or description of, the earth. Its practitioners since the days of Homer have written about the lands and peoples of the habited world, the Greek *ecumene*—and speculated about lands and peoples beyond the range of human knowledge.

The most complete definitions of the scope and purpose of geography now extant are contained in the works of Ptolemy and Strabo.[1] We conclude from their works that geography is concerned with the location and interconnections of places on the earth as a whole. Chorography deals with the integral place of all parts of a given whole as limbs are to the human body, and topography deals with the place of a discrete unit, series, or group of units. Those terms are based on the Greek word-roots *ge*, *chora*, and *topos*, meaning earth, district and place. These word-roots were redefined in the writings of Varenius in the seventeenth and Kant in the eighteenth centuries. The idea behind this conceptual framework is what Strabo described as the 'natural attributes of place' within a framework of relation to other places on the earth's surface. No matter how sophisticated modern concepts and problems become,

---

[1] F. Lukerman, 'The Concept of Location in Classical Geography,' *Annals Assn. of Amer. Geogrs.*, 51 (1961), pp. 194–210.

3

this is the basic and unique concepts of all geographical ideas, and it remains firmly fixed in popular parlance to this day.

The classical view of location has a double meaning—astronomical position and terrestrial position. The view of the earth as a circular land mass surrounded by a river of ocean was held by such men as Anaximander and Hecataeus of Miletus. It was denied by Herodotus and Aristotle as mythical, since it was not based on either observation or experience. Location was concerned only with the known earth, the *ecumene*. Beyond it lay the unknown, and they indulged in no speculation and certainly did not attempt to map what might be there.

Cosmography, on the other hand, considered the earth's properties that were derived from its being as a body in space. The shape, size, and divisions of the earth were thus the content of cosmography, though Strabo pointed out that cosmographical ideas in reference to the earth were basic to geographical description of its surface and the properties and interrelation of places. The division of the earth into habitable and uninhabitable sectors (whether based on heat or cold) was a cosmographical matter. Similarly, the division of the earth into five parallel zones was a cosmographical concern. This idea probably goes back to the fifth century B.C. in Greece and obviously must have been based on the idea of a spherical earth, though this concept was based on philosophical argument rather than observation. It remained to Aristotle in the fourth century B.C. to prove this sphericity on the basis of observations and deduction and for others later to calculate its dimensions. These zones, divided by the arctic and tropical circles into torrid, temperate, and frigid, were based on the projection of celestial properties on the surface of the terrestrial sphere long before the earth was actually circumnavigated.

APIAN AND MUNSTER

From Ptolemy and Strabo we now leap across a period of more than fifteen hundred years. Geography as an organized body of knowledge made little progress in the so-called dark and middle ages. During the Renaissance classical ideas were revived and, as far as geographical knowledge was concerned, were only gradually discarded or rectified. For example, Ptolemy's map, and his calculations of latitudes and longitudes of places, affected map makers and distorted the portrayal of coastlines until the eighteenth century. Indeed, explorers were still searching for the Mountains of the Moon

4

in the nineteenth century, and ideas about the distribution of land and water in the southern seas were dominated by the Ptolemaic notion of a southern continent until Captain Cook's voyages in the eighteenth century.

Brief reference must be made to Peter Apian (Petrus Apianus) and Sebastian Münster who produced two geographical works at the beginning of the Age of Discovery, which for a hundred years after their publication in the sixteenth century were the chief standard theoretical works; the one for its popular exposition of astronomy and mathematical geography and the other for its descriptive geography modelled on Strabo.

Peter Apian, who was born in Saxony in 1495, was an astronomer and cartographer. In addition to making maps and globes, he published two works, *Astronomicon caesareum*, an astronomical treatise, and *Cosmographicus liber*, which he modelled on the *Cosmographiae introductio* of Martin Waldseemüller. The *Cosmographicus liber* was first published in 1524. The book in its original form deals almost exclusively with those aspects of geometry and astronomy which are essential to geography. Latitudes and longitudes for numerous places are given, and what descriptive geography it contains was subsequently appended by Gemma Frisius who follows closely the methods and descriptions of Ptolemy.

The Ptolemaic distinction between geography and chorography is clearly stated. The earth is shown at the centre of the universe; the sun and planets revolving round it. The earth is divided (in accordance with the Aristotelian tradition) into five zones: torrid, temperate (between the tropics and polar circles), and frigid. *Climata* are defined as the spaces between parallels of latitude at intervals of a difference of half an hour in the length of their longest day, and each *climata* is named after the principal feature in it, such as a town, a river, or a mountain range. The lands are of four forms, namely islands, peninsulas, isthmuses, and continents, each illustrated by a simple diagram. Then follow diagrams of hands and feet to serve as a basis for linear measurement. A short note is given on each continent, and finally there is a long list of towns for each country, with their latitudes and longitudes taken from Johannes Schöner and Ptolemy.

Sebastian Münster is the best representative of the German geographers in the sixteenth century. Born at Ingelheim near Mainz in 1489, he studied at Heidelberg and Vienna, and in 1536 he was appointed to a chair of Hebrew at Basle, where he remained until his death in 1552. Münster made important contributions to cartography. He attempted to improve cartographic methods by using a small compass, the forerunner of the prismatic compass, for a simple triangulation survey of a small area round Heidelberg.

Münster published his edition of Ptolemy at Basle in 1540 and during the next twelve years four re-editions appeared. This work was followed in 1544 by the *Cosmographia universalis* which is a compilation from many contemporary authorities rather than a carefully arranged treatise. History and genealogical tables take up a large part of the book, and mathematical and physical geography are almost excluded.

Münster takes for granted that the earth is spherical and he declares that the earth's crust suffers changes through floods and the work of rivers. His knowledge of floods in Holland caused him to declare that many lands have been flooded since the Deluge, and that mountains and valleys have been formed by rivers where the land was formerly flat. He also mentions earthquakes, the 'central fire', the character of rocks, the nature and distribution of metals, and methods of mining. The book deals primarily with human and political geography of Germany on a chorographic basis. The last part of the book deals with Asia, Africa, and the New World. All the material here is second hand, and the descriptions are much inferior to the section on Europe.

Münster was a keen observer and a good writer and his work was the standard text for more than a hundred years. 'So completely did the volume resulting from the insight, learning, and energy of Münster meet the demand of the time, and so thoroughly did it establish itself, that in enlarged form it remained in use until after 1650, going through forty-six editions and appearing in six languages.'[1]

VARENIUS, CLUVERIUS, AND KANT

The successor to *Cosmographia universalis* was Bernhard Varenius' *Geographia generalis* which appeared in 1650. This work was the first which sought to combine general, mathematical, and physical geography and chorography.[2]

[1] Allan H. Gilbert, 'Pierre Davity: His Geography and its use by Milton', *Geog. Rev.*, 7 (1919), pp. 322–38; reference on p. 325.

[2] An appraisal of the works of Varenius was presented by J. N. L. Baker in 1955 in the *Transactions of the Institute of British Geographers*. This is reproduced in *The History of Geography*, Oxford and New York, 1963, pp. 105–118. Varenius' *Descriptio regni Japonicae et Siam* was published in 1649 at Amsterdam and in 1673 in London. The *Geographia generalis* was first published in 1650. Two editions, edited by Sir Isaac Newton, were printed in Latin in Cambridge in 1672

From Strabo to Kant

Bernhard Varenius was born in 1622 at Hitzacker, a small town on the Elbe near Hamburg. In 1640 he entered the gymnasium at Hamburg and studied philosophy, mathematics, and physics. After three years he went to the university at Königsberg to study medicine but, dissatisfied with the teaching, he moved to Leiden to pursue the same work. In 1649 he published his first book on the geography and history of Japan which is an excellent descriptive compilation in view of the limited material at his disposal. This was immediately followed by a companion volume on the religion of Japan. In August, 1650, he published *Geographia generalis*, which was written between the autumn of 1649 and spring of 1650. The work should undoubtedly have been followed by a second volume, but it remained unfinished owing to his premature death, at the age of twenty-eight, in 1650.

The full title of an early translation of Varenius' geography is *A Complete System of General Geography: Explaining the Nature and Properties of the Earth . . . Originally Written in Latin by Bernhard Varenius, M.D., Since Improved and Illustrated by Sir Isaac Newton and Dr. Jurin; And now Translated into English . . . by Mr. Dugdale. The Whole Revised and Corrected by Peter Shaw.* A long subtitle describes 'The Nature and Properties of the Earth' as follows:

Its Figure, Magnitude, Motions, Situation, Contents, and Division into Land and Water, Mountains, Woods, Deserts, Lakes, Rivers, etc.

With particular Accounts of the different Appearance of the Heavens in different Countries; the Seasons of the Year over all the Globe; the Tides of the Sea; Bays, Capes, Islands, Rocks, Sand-Banks, and Shelves. The State of the Atmosphere; the Nature of Exhalations; Storms, Tornados, etc.

The Origin of Springs, Mineral Waters, Burning Mountains, Mines, etc.

The Uses and Making of Maps, Globes, and Sea Charts.

The Foundations of *Dialling*; the Art of *Measuring Heights* and *Distances*; the Art of Ship-Building, *Navigation*, and the Ways of *Finding the* Longitude at Sea.

Although Varenius was hindered by lack of material, his ideas

---

and 1681 respectively. The first English translation by Blome dates from 1693. Another edition with J. Jurin as editor appeared in 1712. The Jurin edition was further edited by Dugdale and Shaw and appeared in 1736. It contained amendments and additional footnotes. The fourth, and last, English-language edition was published in 1765. This is the edition quoted here as indicative of the level of geographical knowledge in the middle of the eighteenth century.

were far in advance of the knowledge of his time. He defines geography as follows:

Geography is that part of *mixed Mathematics*, which explains the State of the Earth, and of its Parts, depending on Quantity, *viz.* its Figure, Place, Magnitude, and Motion, with the Celestial Appearances, etc. By some it is taken in too limited a Sense, for a bare Description of the several Countries; and by others too extensively, who along with such a Description would have their Political Constitution.

Geography is divided into two divisions, general or universal and special or particular.

We call that *Universal Geography* which considers the whole Earth in general, and explains its Properties without regard to particular Countries: But *Special* or *Particular Geography*, describes the Constitution and Situation of each single country by itself which is twofold, *viz. Chorographical*, which describes Countries of a considerable Extent; or *Topographical*, which gives a View of some place or small Tract of the Earth.

General or universal Geography is divided into three parts: the absolute part which deals with the form, dimensions, and position of the earth, as well as the distribution of land and water, mountains, woods, deserts, and hydrography and the atmosphere; the relative part which deals with the 'Appearances and Accidents that happen to (the earth) from Celestial Causes', that is latitude, climatic zones, longitude, and so forth; and the comparative part which contains 'an explication of those Properties, which arise from comparing different Parts of the Earth together'.

Special Geography, though not dealt with in detail, is outlined under three 'particulars or occurrences'. The 'Celestial Properties are such as affect us by reason of the apparent Motion of the Sun and Stars'. Terrestrial properties are those which 'are observed in the Face of every Country' and include location, shape, magnitude, topography, hydrography, vegetation, fertility, minerals and soils, animals, and longitude. Human properties concern 'the inhabitants of the place' and again fall into ten categories, as follows:

1. Their Stature, Shape, Colour, and the length of their Lives; their Origin, Meat, and Drink. 2. Their Arts, and the Profits which arise from them; with the Merchandise and Wares they barter with one another. 3. Their Virtues and Vices, Learning, Capacities, and Schools. 4. Their Ceremonies at Births, Marriages, and Funerals. 5. The Language which the Inhabitants use. 6. Their

Political Government. 7. Their Religion and Church Government. 8. Their Cities and famous Places. 9. Their remarkable Histories. 10. Their famous Men, Artificers, and the Inventions, of the Natives.

Varenius complained that special geography was always taught at the expense of general, and on this account he argued that geography scarcely merited the dignity of a science. In special geography features should be explained in terms of general laws, so as to make local geography logical and intelligible. It is therefore not surprising that in his book on Japan, which is special geography, there is a close correspondence of topics with the divisions of general geography.

A few years before the publication of Varenius' book, two other works were published, one by the German Philip Cluverius (Cluver)[1] and the other by Nathaniel Carpenter, an Englishman.[2] I confine attention to the first.

Cluverius (1580–1622) who travelled widely in Europe approached geography through the classics and history. His general work, *Introductio in Universam Geographiam* was published posthumously in 1624. This work preserves the traditional distinction between geography and chorography. Of the six books in the volume, only one deals with the earth in general, while the remaining five contain brief descriptions of countries in which the human and historical elements are stressed. The general geography is decidedly inferior to that of Varenius. Cluverius does not know the views of Copernicus; for him, the earth remains the centre of the universe. His mathematical geography and astronomy show no advance on the work of Apian written a century earlier, and physical geography is limited to the distribution of land and water. Cluverius thus concentrates on the description of countries, which are described generally in terms of their name, extent, and nature of the land and its products and the ancient and modern political divisions, ethnography, and topography.

The Ptolemaic definitions of geography, chorography, and topography are given. There follow statements of the astronomical zones,

---

[1] P. Cluverius, *Introductio in Universam Geographiam*. The bulk of this work was translated for the present writer from the Latin version in the British Museum by the late M. T. Smiley, Professor of Latin and Greek at University College, London, in 1932. This is the first opportunity I have had to express in print the thanks of a young man to a senior and respected colleague.

[2] J. N. L. Baker, 'Nathaniel Carpenter and English Geography in the seventeenth Century', *Geographical Journal*, 71, (1928) pp. 261–71, reprinted in *The History of Geography*, Oxford, 1963.

great circles, zodiac, meridians, minor circles, parallels and climate, cardinal points, divisions into 360 degrees and measurements of distances, the ocean and its divisions, the great inlets of the ocean, the inner sea (that is, the Mediterranean), navigation, and a summary division of the whole world.

The method of description for a country may be illustrated by the British Isles, which are briefly portrayed as to location and extent. Tribal units, countries, and the inhabitants (mainly referring to the Roman occupation) are described, and this is followed by short sections on England, Scotland, and Ireland. The section on Ireland reads as follows:

> What peoples were the first to inhabit it is uncertain, except that like the rest of the Britons, they were of Celtic stock. The Brigantes, Cauci, and Menapii, whom Ptolemy mentions as in it, very probably crossed thither from Gaul, Britain, and Germany. Today its chief division is into four districts. There are thirty-three counties in the whole kingdom. It would be more correct to say that the Irish live in small towns than in cities.
>
> It has such an abundance of grass, that is rich and sweet, that cattle take but a small part of a day to fill themselves; and unless they are kept from the fodder, they go on grazing and burst asunder. Its population is uncouth, and a stranger to all virtues, to a greater degree than other races.

The works of Cluverius and Varenius were standard until the middle of the eighteenth century and bring us to the threshold of the development of modern geography as a discipline.

Immanuel Kant (1724–1804), the philosopher, lectured on physical geography in the University of Königsberg from 1765, and his lectures were subsequently published.[1] In his view the human element was an integral part of the subject matter of geography. Kant divided the communication of experience between persons into two branches, narrative or historical, and descriptive or geographical;

---

[1] D. F. T. Rink, *Kant's Physische Geographie*, 1802. Kant's *Gesammelte Schriften*, Bank IX, 1802 Berlin, 1923. G. Gerland, 'Immanuel Kant, seine geographischen und anthropologischen Arbeiten,' *Kant Studien*, 19 (1905), pp. 417–547. E. Adickes, *Kant's Ansichten uber Geschichte und Bau der Erde*, Tübingen, 1911; *Untersuchungen zu Kant's physischer Geographie*, Tübingen, 1911; and *Ein neu aufgefundenes Kollegheft nach Kant's Vorlesung uber physische Geographie*, Tübingen, 1913. The works by Adickes are reviewed by O. Schlüter in the *Geog. Zeit.*, 19 (1913) and 20 (1914). See also an important discussion of Kant's geographical ideas in Hartshorne's *The Nature of Geography*, especially pp. 38–44. Also G. Taylor (ed.) *Geography in the Twentieth Century*, 1951, containing G. Tatham's essay on pp. 38–42.

and he regarded both history and geography as descriptions, the former in time and the latter in space. He claimed physical geography to be 'a summary of nature', the basis not only of history but also of 'all the other possible geographies'. Five of the other geographies he identified as part of physical geography are mathematical (the form, size, and movement of the earth and its place in the solar system), moral (the customs and character of man in relation to environment), political, mercantile (commercial), and theological (the distribution of religions). Physical geography thus embraced the outer physical world, the earth's surface, and its cover of life forms of plants, animals, and man and his works.

Kant followed essentially the Ptolemaic definitions of geography. However, there was an essential difference between his work and that of the classical scholars. The latter laid primary emphasis on the areal divisions of the earth and the business of systematically describing their distinctive content. Kant, on the other hand, was not so much concerned with composite terrestrial units of different orders, but with the orderly investigation of particular areally differentiated phenomena. Both are integral aspects of the areal differentiations of the earth's surface, but out of the difference of emphasis that is reflected in the works of Ptolemy and Strabo on the one hand and Kant and Varenius on the other emerges differences of approach that run throughout the development of modern geography.

These differences of approach have been recently described as the theoretical (deductive) or nomothetic approach and the empirical (descriptive) approach. The theoretical approach seeks to establish theories relevant to the location and interrelations of places and to establish laws and make deductions on the basis of the laws. The empirical or ideographic approach places primary emphasis on the description of particular groupings of nations and people in terms of lands, seas, countries, and places. It does not seek to develop laws but to find out how phenomena account for the *genus loci*, the character of a place and its relations with other places. These are the two basic approaches and traditions in all geographical inquiry and their contrast and conflict have become more marked and difficult to bridge as knowledge of the surface of the earth has increased.

NATURE V. MAN: THE WORK OF BUFFON

At the opening of the nineteenth century, there was concern with the distinctive concepts and procedures of the study of geography as an organized body of knowledge. Three matters need special attention:

first, the modes of collation and interpretation of the data relevant to the earth's surface and the life forms upon it, taking the colossal work of Count Buffon as a major indicator; second, the problems, both theoretical and applied, of defining areal units as a basis for description of areas; and, third, the content of standard geographies published about this time that reveal not only the scope and methods of geographical writing, but also the status of knowledge of the earth's surface and its differentiated parts.

By the end of the eighteenth century there was gradually emerging a theory which favoured the slow operation of natural laws as opposed to the cataclysmic events occuring through divine intervention. At that time both beliefs were largely theoretical. In the early nineteenth century, however, the evidence in favour of the slow operation of natural laws was collected and their overwhelming impact came in the 1860's.

The problem of developing scientific methods for the study of the life forms also received attention in the late eighteenth century. The passion for the scientific collection of specimens on voyages of exploration resulted in remarkable assemblages such as the flora brought back by the botanists on Cook's voyages of discovery. The task of classification and description of flora and fauna was also undertaken. Leibnitz and Buffon tried to arrange all life forms in a scale from the simplest organisms to the most complex. Linnaeus, on the other hand, arranged all life forms into taxonomic groupings.

The question of the nature and growth of civilization drew the attention of many thinkers. Theories of the progress of mankind involved some speculation on the role of the physical environment as a stimulant or deterrent. Buffon and others were concerned with the question of the unity of the human race and the manner of its dispersion over the entire world. This belief in the unity of the human race required explanation of the diversity of man and his uneven distribution over the earth and asked questions about the relation between population density and productivity.

All these trends are clearly reflected in the voluminous *Histoire naturelle, générale et particulière* by Georges-Louis Leclerc, Comte de Buffon (1707–88). The *Histoire naturelle* is in forty-four volumes published between 1749 and 1804. It was completed after Buffon's death by La Cépède.

Clarence Glacken[1] has evaluated Buffon's work as follows: 'The *Histoire naturelle* satisfied a hunger for concrete and detailed knowledge that was not dependent on mathematics or Cartesian deductive

[1] C. J. Glacken, 'Count Buffon on Cultural Changes of the Physical Environment,' *Annals Ass. Amer. Geogs.*, 50 (1960), pp. 1–21.

reasoning, but on study, travel, observation, and description.' This offers a background for understanding the philosophical views of scientific developments in this remarkable era, that was marked by a prodigious enthusiasm for exploration, travel, and the collection of data about the earth, its plants and fauna, and its primitive societies.

While covering the whole realm of nature, Buffon makes repeated reference to the relations between man and his environment. He refers to the changes which man has made in his natural environment, and particularly to the transformations which have accompanied the growth and expansion of civilization and the migration and dispersion of human beings and their domesticated plants and animals throughout the habitable parts of the earth. This is the main theme of *Des époques de la nature*, whose title refers to the seven periods of the earth's history: 'the formation of the earth and the planets, the consolidation of the rock in the interior of the earth, the invasion of the continents by the seas, the retreat of the seas and the beginning of volcanic activity, the north as the habitat of elephants and other animals of the south, the separation of the continents, and the power of man aiding that of nature.' It is in this last phase that Buffon draws attention to the relationship between man and nature. He envisages Central Asia between 40° and 55° as the seat of the earliest civilizations, which had developed thirty centuries before in an environment of pleasant climate and fertile soil, sheltered from floods and free from earthquakes. There civilized man first attained knowledge, science, and power. But this civilization was destroyed by people from the north who were driven out of their homeland by overpopulation. Although much of the early civilization was lost, agriculture and building techniques survived and were diffused into China, Atlantis, Egypt, Rome and Europe. In the succeeding centuries man continued to progress, and Buffon was able to conclude:

> The entire face of the Earth bears today the stamp of the power of man, which although subordinate to that of Nature, often has done more than she, or at least has so marvellously aided her, that it is with the help of our hands that she has developed to her full extent and that she has gradually arrived at the point of perfection and of magnificence in which we see her today.[1]

Buffon clearly believed in the creative power of man. As an agronomist, he wrote much about forest clearance and the need for conservation. He developed the theme that by changes of the landscape—forest clearance and drainage in particular—the climate became warmer. He speculated on climatic change and suggested

[1] C. J. Glacken, *op cit*. p. 10.

that the removal of forests and the drainage of marshes might lead to temperature increases. On the other hand, he urged the conservation of forests. Glacken summarizes Buffon's position by saying 'large areas inimical to man had to be cleared, . . . but once societies were established on them, the forests were resources which had to be treated with care and foresight.'

Buffon was one of the early writers on soils. These he grouped into clays, calcareous earths, and vegetable earths, the last falling into two forms, the *terrean* (leaf mould) and the *limon*, which is the residue in the decomposition of the terrean.

He also wrote at length on domesticated animals and plants and made frequent shrewd comments on landscape, especially the contrasts between the habited and uninhabited land. Glacken presents Buffon's viewpoint:[1]

> Among countries inhabited for a long time, there are few woods, lakes, or marshes, but they have many heaths and shrubs (meaning no doubt that heaths and shrub take over deforested and barren mountain tops). Men destroy, drain, and in time give a totally different appearance to the face of the earth.

THE REGIONAL CONCEPT AROUND 1800: PURE GEOGRAPHY[2]

Geographies before the mid-eighteenth century were utilitarian encyclopedic compilations, without orderly presentation or general principles of areal distribution and with reference only to existing political units. Europe, the best known and most fully described part of the world, was at that time divided into a crazy interlocking mosaic of political divisions with impermanent boundaries which could not possibly be used as a rational basis for description.

The new ideas of geographic description were primarily due to German writers. This was certainly owing to the fact that there were some four hundred political divisions in the German-speaking part of Europe. At the same time these German writers were trying to decide

---

[1] C. J. Clacken, *op cit.* p. 19.

[2] The main reference in English to the period around 1800 is Hartshorne's *The Nature of Geography*, pp. 211–24 (35–48). Ernst Plewe discusses the life and work of A. F. Büsching in an essay in the Hermann Lautensach Festschrift, *Stuttgarter Geog. Studien*, Band 69, ed. H. Wilhelmy, 1957. A main source in Germany is E. Wisotski *Zeitströmungen in der Geographie*, Leipzig, 1897. There is also a brief, but useful, discussion and bibliography in K. Bürger, *Der Landschaftsbegriff: Ein Beitrag zur geographischen Auffassung*, Dresdner Geog. Studien, Heft 7 (1935).

just what the historical term *Deutschland* meant—politically, culturally, and linguistically—as a geographical term. Out of the resulting difficulties of descriptive presentation, there emerged interest in *reine Geographie* or pure geography. This meant the use of natural, or physical, land units instead of political units as a basis for description of both land and people. The two kinds of study were called *Länderkunde* and *Staatenkunde* respectively.

Anton Friedrich Busching wrote the *Neue Erdbeschreibung* (eleven parts; 1754–92) in which he sought to provide 'a description of the known surface of the earth', but he used political units as his regional base, and his presentation did not differ from the many other works based on the pattern of Strabo.

A new kind of natural division of both lands and seas, however, was proclaimed by the French scholar, Philippe Buache in 1756 in his *Essai de géographie physique*. Basing his theories on a 1737 bathymetric survey of the English Channel, he postulated that the earth was divided into drainage basins by mountain ranges. This speculation was based on limited knowledge of Europe and dismal ignorance of both the surface configuration of the continents and of the ocean floors. Buache recognized three categories of mountains: the highest ranges which are oriented from west to east or north to south; the medium size ranges that are branches of the highest and which separate the main river basins from each other; and the low ranges which are branches of the medium and the source of shorter streams. In similar fashion, he envisaged the seas as divided into three oceans and two arctic seas (the southern sea being purely hypothetical and, of course, incorrect). Each of the three main oceans (Atlantic, Pacific, and Indian) is divided into basins by underwater ridges.

This idea of the natural division of the earth long remained in favour. It was taken up by a German, Johann Christoph Gatterer, in his *Abriss der Erdbeschreibung* in 1775. However, the bulk of his description is still based on political units. Gatterer's geography embraced countries and their boundaries (*Länderkunde*), political states (*Staatenkunde*), and human and ethnic groups (*Völkerunde*). He divided Europe into west and east. Western Europe, called *germanische Europa*, he divided by the Pyrenees and the Alps, and by the Baltic Sea and the *Britische Meer*, into four divisions. These divisions are the Pyrenean peninsula (Spain and Portugal); the Alpine Lands (West Alps and France, South Alps and Italy, North Alps and Switzerland, Germany, and Netherlands); the British Isles; the Baltic Lands which included all of Scandinavia and Prussia. He included all the remaining lands of Europe in eastern

Europe—Russia, Hungary, and Turkey. These groupings of states Gatterer claims to have based on 'natural divisions', but the boundaries remain those of political states.

We may also note here that Gatterer defined geography as 'the description of the earth as a portrayal of the earth's surface, what it is and what Man has sought to make of it; a description of the heavens, the earth, and their influences upon each other.' Of special interest to him was: 'the practical and historical description of the earth that is concerned with the earth as the home of man'.

The idea of the natural division of the earth's surface was developed by others in this period. H. G. Hommeyer wrote a work on the military geography of the European States (*Beiträge Zur Militärgeographie der europäischen Staaten*) in 1805. He complained that his contemporaries treated topographic description on the basis of political units. Hommeyer argued that political units were only an incidental part of the training and need of the military officer, whose primary concern was with the natural conditions of the earth as the theatre of war.

Among these writers we find the beginnings of the concept not only of major 'natural divisions', but also of a hierarchy of divisions within the major units. Hommeyer, for example, envisaged four orders of geographical unit. The locality or *Ort* is the smallest geographical unit. The district or *Gegend* is the land area visible from a high vantage point. The region or *Landschaft* is 'the area visible from a very high vantage point', and is 'a number of districts that are contiguous and clearly separated from neighbouring groups of districts, chiefly demarcated by mountains and forests'. The fourth and largest unit is the *Land* which is a part of the earth's surface bounded by the general arrangement of drainage, and is of such a size that 'the localities and the districts and the limited changes associated with them have little influence on the location or the orientation of the ranges of the *Land*'.

Other German writers of this period also believed that geographical study should have its basis in natural rather than in political units. We should note in particular the work of Johann August Zeune, the predecessor of Ritter in the chair of geography at Berlin. In his book, *Gea: Versuch einer wissenschaftlichen Erdbeschreibung*, which was published in 1808, Zeune discarded the changing 'mosaic of the political States' and turned to the lasting base of the physical earth. He referred to the work of Gatterer, but his view apparently *excluded the work of man*. He divided Europe into its major divisions. Southern Europe falls into three units—the Pyrenean, Alpine, and Balkan peninsulas. *Mitteleuropa* covers the

16

Carpathian lands, Hercynian land, and Cevennes. The Carpathian lands embrace Hungary and 'other Slav lands' bounded by the Balkans to the south and the Carpathians and the Dniester to the north. Hercynian land included Germany and Denmark, as well as parts of Switzerland and Holland. Its natural boundaries were the sea to the north, the Alps to the south, the Rhine to the west, and the Oder to the east. This is, in fact, the historical Deutschland. The Cevennes embraced France, plus parts of Switzerland and Holland, and had its boundaries in the Pyrenees, Alps (as far as the St. Gotthard Pass), and the Rhine to its mouth. North Europe also has three parts—the North Sea islands (Great Britain), the East Sea peninsula (Sweden with Finland and Norway), and Russia, bounded by the Urals to the east, the Dniester, the Carpathians and the Oder river to the southwest, the Baltic, Gulf of Finland, Lake Ladoga, and White Sea to the northwest. He writes that these 'land units' are mainly drainage units bounded by ranges, and that 'one finds a tendency of Nature to create similar plant and animal life, and indeed even similar human and folk traits . . . Thus the mountain ranges are natural plant, animal, and human divides.' This is an explicit form of environmental determinism that did not find general acceptance among his contemporaries.

Similar ideas were expressed by A. L. Bucher. In a work dated 1812, (*Betrachtungen über die Geographie und über ihr Verhältnis zur Geschichte und Statistik*) he recognized three aspects of geographic description, the study of states (*Staatenkunde*), of peoples (*Völkerkunde*), and of lands (*Länderkunde or Spezialgeographie*) which dealt with 'the natural content of areas'. Bucher defines four orders of 'geographical land units'. These are the continents; the larger divisions (*Grossgebiete*) of the continents; the 'natural lands', such as the northern Alpine Lands containing the Rhine, Weser, and Elbe Rivers; and the subdivision of the natural lands, such as the East Jara (Upper Rhine and tributaries, West Sudetenland (Bohemia), and East Harzland (Elbe drainage area). He writes of the 'interaction of Nature and Man' and says that these divisions 'in their total physical character have more similarities with each other than with the next nearest districts that are more closely related to another division'.

In all these works the usual arrangement was into mathematical, physical, and political sections. Few writers made a distinction between a geography of non-human features and a geography of man. Physical geography and man were associated in area and in causal relationships to each other. There is also apparent the concept of the interdependence of spatially arranged phenomena so as to form 'organic wholes'. The idea of a composite whole, however, though put forward by Bucher in 1812, met with opposition, and Bucher himself in 1827 (*Von den Hindernissen, welche der Einführung eines besseren Ganges beym Vortrage der Erdkunde auf Schulen Im Wege stehen*) argued against it. He then considered that areas should be defined for specific purposes. The essential work of geography, he argued, was concerned with the systematic study of individual categories of phenomena, each in its relations to the earth. Hartshorne rightly concludes: (*Nature of Geography*, p. 222) 'the arguments with which Bucher attacked the concepts of natural boundaries and natural regions leave little for the present-day writer to add'. We shall return to this theme. It is one of the persistent problems of geographic work and it is important to learn that it was so clearly envisaged and stated a hundred and fifty years ago.

Among the late eighteenth-century geographers a special place must be given to the two Forsters, Johann Reinhold (1729–98) and Johann Georg Adam (1754–95), father and son, who accompanied Cook on his second world voyage. J. R. Forster's *Observations Made during a Voyage Round The World*, published in England in 1778, are arranged under six headings—earth and lands; water and the ocean; atmosphere; changes of the globe; organic bodies; and human species. He recognized the ties between man and his environment and the mobility of peoples. Oscar Peschel described him as 'the first traveller to give a physical survey of the section of the world he had seen (South Seas) and the first to perform the highest function of a geographer, namely, that of scientific comparisons'. Ernst Plewe has recently acclaimed him as 'the first great German methodological geographer in the modern sense'.

Georg Forster, his son, is known better than the father, since he translated his father's book into German (1778). He met Humboldt at Göttingen in 1789 and then began a friendship that had a profound influence on Humboldt. Peschel wrote that Forster 'was the first writer to awaken love and feeling for the beauty and scenery' and Plewe writes that in a description of the lower Rhine lands (written after a trip with Humboldt) 'he founded more securely his father's method and prepared the way for a systematic development of regional geography'.

THE FIRST UNIVERSAL GEOGRAPHY:
MALTE-BRUN

The definitive geographic compilations of this period, in addition to the German works quoted above, were the work of Pinkerton (Britain), Malte-Brun (France) and Jedidiah Morse (America). The second only will be briefly reviewed here.

Conrad Malte-Brun (1775–1826) was born in Denmark, banished from that country in 1800 and settled in France. His *Précis de la Géographie Universelle* contained six volumes and was published between 1810 and 1829 (though the work was completely by colleagues after his death). This great work is entitled, in its English translation (Philadelphia, 1827–32), *Universal Geography or a Description of All Parts of the World on a New Plan according to the great natural divisions of the globe, accompanied with analytical, synoptical and elementary tables*. The work begins with 'the general theory of geography', which consists of 'its mathematical, physical and political principles'. It then passes to 'the leading features of nature' and then to 'animals, the plants, all the beings that are nourished in the exhaustless bosom of the earth' in 'their native regions'. This introduction concludes with 'man in his natural and his political condition', meaning races, languages, beliefs and 'the laws which mark the progress of civilisation'. The first half of the first volume deals with the 'theory of geography'. The second half and the next five volumes contain general sections on the world's major divisions. These divisions are selected on a basis that is essentially the same as the procedure of Strabo, two thousand years earlier. Description is on the basis of countries, variously defined. For example, Asia is divided as follows—Caucasian Countries, Turkey-in-Asia (Asia Minor and coasts of the Black Sea, Armenia, Mesopotamia, Syria and Palestine), Arabia, Persia, Caspian Sea, Afghanistan, Tartary, Siberia, Central Asia, Manchuria and Korea, Japanese Islands, China, Tibet, Industan, Indo-China, and Oceanica. Malte-Brun writes as follows about his method of description.

> When we mean to give a description of an extensive country in detail, it presents itself in two different points of view, which have two corresponding modes of subdivision. It may be divided into governments, provinces and districts. It may also be divided according to the nations which inhabit it. The one of these methods is that of chorography; the other, that of ethnography. We usually begin with the first.

The volume on Europe first deals with 'physical geography'. This embraces seas, lakes, rivers, mountains, climate (including isothermal lines), and animals and plants. Then follows what is called 'political geography' which embraces nations, languages, religions, political divisions, governments, and population.

In order to get an idea of the method of treatment of smaller areas and to compare it with the presentation of the 'universal geographies' of Reclus and Vidal that followed it, and which will be discussed later, we may note here the treatment of France. This begins with a peculiar and brief 'historical notice'. This is followed by a section on 'physical geography' beginning with mountains, rivers, capes and bays, and ending with 'insects, horses, oxen, pigs, poultry, importations'. For purposes of further detailed description, the country is divided into five parts or 'regions' each of which presumably is supposed to have certain common characteristics. This procedure it is stated, is 'the one by which its chorography may be most conveniently explained'.

The work of Malte-Brun, in spite of its glaring inadequacies, is a landmark in the history of geographic description, for it completed what Ritter in the same period failed to achieve—a description of the lands and peoples of the earth. Unlike the work of Ritter, however, it shows little, if any, advancement on the conceptual framework and procedure of Strabo who preceded him by almost two thousand years. The tradition and need, however, were obviously strong and lasting. What was needed was a more scientific approach and more reliable data and maps for the rational explanation of the areal variations of lands and peoples over the face of the earth. This became a central concern of the development of modern geography and is reflected in the second work of this kind and with the same title by Élisée Reclus in the third quarter of the nineteenth century; by the third work, brilliantly conceived by the founder of modern geography in France, Vidal de la Blache, and brought to fruition by his immediate followers in the second half of the twentieth century; and by the handbook of geographical science published under the stimulus of Alfred Hettner in Germany. These are widely acknowledged contributions to modern scholarship.

It is astonishing to record the continuity of the classical tradition of geographic writing from Strabo to Malte-Brun. The problems of theory and compilation at the beginning of the nineteenth century were in all essentials those of the classical Greek scholars nearly two thousand years earlier. The world was still mainly unexplored and unmapped, (though the coastline outlines were correctly mapped at the end of the eighteenth century, notably by French cartographers). There was an appalling lack of data about the lands, resources and peoples in individual places, and an almost complete absence of scientific methods of generalising such data on maps. Without such data and the cartographic means of handling them progress in geographical thought was impossible. The Classical

From Strabo to Kant

Tradition, still strong around 1800, and continued by the founders of modern geography in the nineteenth century, had its two greatest exponents in Ritter and Humboldt. Imbued by the philosophy of their day (as represented in the great work of Count Buffon), they stand as the culmination of the classical tradition.

In spite of a woeful lack of data which their theory and practice needed, they sought to collect their own, and to establish more scientific guide-lines than those in general use. Their guide-lines, though based essentially on the concepts of the pre-evolutionary era, have served and still serve (by general acknowledgement) as a sound theoretical framework for the development of geography as a modern field of study with less pretentious claims, and a more clearly defined field of objectives, methods and expertise.

The theory of the classical founders outran the facts. The task of scholars from Humboldt and Ritter to our day has been to collect and collate facts of lands and peoples and to find methods of measuring and mapping them. Only then could substantial advances be made in the realm of theory regarding the variations of land and peoples over the surface of the earth.

Further References

There are many studies of classical and medieval geography, mainly by linguistic (classical) scholars, who have little real understanding of the objectives of the writers. Two works by geographers therefore need special mention. These are Chr. van Paassen, *The Classical Tradition in Geography*, Groningen, 1957 and G. H. T. Kimble, *Geography in the Middle Ages*, London, 1938. Reference should also be made to the publications of the Research Series of the American Geographical Society, New York.

2

Alexander von Humboldt (1769-1859)

Alexander von Humboldt and Carl Ritter (Ch. 3), both outstanding scholars in the first half of the nineteenth century, are recognised in both Germany and France as the founders of modern geography. They both died in 1859, the year of the publication of Charles Darwin's *Origin of Species*. Their ideas and purposes, although formulated before the great impact of evolutionary thought, and before the enormous extension of exploration and mapping in the latter half of the century, have continued to serve to this day as guide posts in the field of geography. Discussion here is limited strictly to their contributions to the regional concept, which they both regarded as the core of geography.

Alexander von Humboldt,[1] born in Berlin in 1769, first became a mining engineer after pursuing his studies in geology at Freiburg in Saxony under the famous geologist, A. G. Werner. He soon turned to the career of a natural scientist and explorer, aided by an income inherited after the death of his mother in 1797. On June 4th, 1797, he sailed from La Corunna in Spain, backed by royal permission, to visit the Spanish lands of South America, accompanied by Aimé Bonpland, a French botanist. For five years he travelled across the *Llanos*, sailed up the Orinoco, visited Cuba and Mexico, journeyed extensively in the northern Andes, and climbed many mountain peaks, including Chimborazo, the highest known peak in the world at the time. On returning to France in 1804, he lived in Paris until 1827, during which period he published (at great financial cost) the scientific results of his travels. He settled in Berlin in 1827 with the post of Chamberlain to the King of Prussia. In 1829, at the age of 60, he travelled by carriage into the heart of Siberia at the invitation of the Tsar, to explore the mineral resources. The results of this journey were published in works on the geology, relief and climate of Central Asia. The rest of his life, about 30 years, was devoted to

[1] Helmut de Terra, *The Life and Times of Alexander von Humboldt*, 1769–1859. New York, 1955.

the writing of his best known work, the *Cosmos*. This was completed two days before his death. The Prince Regent ordered a state funeral and the streets of Berlin were draped in black to pay homage to a legendary figure.

Humboldt shared the philosophic views of his era, as expressed by Hegel and Goethe.[1] He shared the concept that envisaged the earth as an inseparable organic whole, all parts of which were mutually interdependent. This was the view of the physical world or the field of *natural philosophy*, the earth and the universe, in so far as accessible to the senses, as opposed to the view of man's inner world or the field of *moral philosophy*. This synthesis he regarded as a harmonious unity, a living whole of all the parts. He derived aesthetic satisfaction through the scientific analysis of the ways in which things and phenomena on the earth's surface depended upon each other (*Zusammenhang*), an idea which runs right through *Cosmos*, and was clearly in his mind as early as 1797 as the *physique du monde*. His field of inquiry was the observable world of phenomena, 'the total impression'—*die erschaute Totalität*.

Humboldt's scientific procedure was empirical and inductive. Empirical knowledge, he often stated, is based on the thoughtful observation of the phenomena revealed to the senses. He collected, classified and interpreted plants, animals, and rocks with respect to their origins and their geographic distribution, and he was the first to draw cross sections of relief in order to show the altitudinal limits and zones of correspondence of the phenomena which he everywhere systematically observed. He also collected masses of measurements of heights, areas, length, and so forth to discover the areal distribution and interrelations of phenomena.

The divisions of knowledge of the natural or physical world, he considered to fall into three groups. First were the phenomena examined taxonomically as to form and content, that is phenomena which were classified according to analagous characteristics (such as botany, zoology, geology). These were the systematic sciences, *histoire naturelle descriptive* or, apparently, *Physiographie*. Second, there were the historical sciences dealing with existing groups of phenomena—the history of the development of animals, plants, and rocks. Such study was concerned with *histoire du globe*, *Erdgeschichte*, or *historia telluris*. Third, there was geognosy or earth science, which he variously defined as *Erdkunde*, *théorie de la terre*, *géographie physique*, *physische Erdbeschreibung*, or physical geography. This is

---

[1] L. Döring, *Wesen und Aufgaben der Geographie bei Alexander von Humboldt*, *Frankfurter Geographische Hefte*, Heft I. 1951, with bibliography, has been freely used in this appraisal.

concerned with the distribution or arrangement of phenomena on the earth's surface. The object of *Erdkunde*, he says, is the study of 'phenomena in their spatial distribution, spatial relationships and interdependence' (*Erscheinungen in ihrer räumlichen Verbreitung, räumlichen Beziehungen und Verknüpfungen*).

Humboldt has not generally been regarded as an outstanding contributor to any *one* field of knowledge. He contributed richly, however, to a variety of fields, but he was always concerned with the *areal associations of natural and organic phenomena*. In the handling of selected natural phenomena, he invariably turned to their distribution and their impact on the spatial arrangement of other phenomena. He made signal contributions to climatology, plant geography, orography, oceanography, cartography, geology, vulcanology and magnetism. Through the study of volcanic processes he came to agree with the Plutonists that many rocks are the result of outpourings of molten material from the earth's interior. But he also brought out features of the *form* of volcanoes, which differ according to their origin, and to their distribution over the earth. He pointed out that distribution of vulcanicity is significant in locating areas of crustal weakness and that volcanoes are associated with areas of earth movement and mountain building. Volcanic features 'in the variety of their forms occur in the most distant parts of the earth,' but there is order and meaning in their distribution.

Climatology[1] is a term used by Humboldt; indeed, he probably coined it. He wrote: 'the term climate means in the general sense all variations of the atmosphere—temperature, humidity, barometric pressure, winds, electric charge, atmospheric purity and degree of visibility.' He saw the measurement of temperature as the basic material from which 'a picture of the distribution of atmospheric warmth' could be obtained. 'Before one can prepare such a system, the facts must be arranged in groups, the measurements taken and assessed ... so that the phenomena of warmth can be brought under empirical laws.' From a compilation of isothermal lines Humboldt was able to demonstrate that near the equator they are parallel to it, but nearer to the poles they spread north over the sea and south over the land and that they reach furthest north on the west coast of continents. He also saw the need for seasonal isotherms. 'Climates of islands and coasts differ from the centre of continents in that the first have milder winters and colder summers than the second.' He speaks of a 'solar climate' as a norm on a uniform globe in which the isotherms would run parallel to the equator. But this is affected by other factors each of which must be analysed and measured.

[1] *Ibid*, pp. 78–83.

Alexander von Humboldt

Plant geography[1] was firmly established by Humboldt on the basis of his massive collection of South American flora. This study reveals 'to our eyes the plant cover, which, here thin, here dense, has been spread by nature over the face of the earth.' He distinguished between the distribution of individual plant species and the numerical association of plants in particular areas. He condemned the numerical listing of flora by continents, since this obscured the distribution of vegetation and its relationship to climatic zones. Instead, he examined the numerical distribution of plants in different isothermal zones. He also pointed out that factors other than climate affected plant distributions.

Types of vegetation and their distribution, he argued, cannot be arrived at simply by adding up the distribution of plants, though certain species can give a key to the composite character of the vegetation in the area in which they occur. He named sixteen indicator plants, including palms, conifers, aloes, and cactus. Forms are the elements of botanical classification. 'The systematic botanist separates a variety of plant groups, which the physiognomist is obliged to bring together.' But such a study is not only physiognomic, 'it must seek laws that determine the physiognomy of Nature in general, the *regional* character of vegetation of the whole of the earth's surface . . . that explains the grouping of contrasted forms in different latitudes and altitudes. It must pursue the causes of the recurrence of all forms of animal and plant life in fixed ever recurring types.' This is the essential base of plant geography, and plant ecology in particular.

Humboldt thus clearly emphasised the physiognomic study of nature. All his work was concerned with the observation of measurable phenomena. He sought to identify homogenous areas however defined. He wrote in his essay on plants (in *Aspects of Nature*) as follows:

> As we recognize in distinct organic beings a determinate physiognomy, and as descriptive botany and zoology, in the restricted sense of the terms, consist in a detailed analysis of animal and vegetable forms, so each region of the earth has a natural physiognomy peculiar to itself.

This is the concept of landscape or scenery based on total visual impression, (*Totaleindruck*).

[1] *Ibid*, pp. 83–8, and especially Humboldt's *Ideen zu einer Geographie der Pflanzen nebst einem Naturgemälde der Tropenländer*, Tübingen, 1807, and *Aspects of Nature* (transl. Sabine, 1849), essay on 'Ideas for a Physiognomy of Plants.' Vol. 2. pp. 13 and infra.

The azure of the sky, the lights and the shadows, the haze resting on the distance, the forms of animals, the succulency of the plants and herbage, the brightness of the foliage, the outline of the mountains, are all elements which determine the total impression characteristic of each district or region . . . But if the characteristic aspect of different portions of the earth's surface depends conjointly on all external phenomena . . . yet it cannot be denied that the vegetable covering with which the whole earth is adorned is the principle element in the impression.

Humboldt included man and his works in this concept of nature and natural areas, but he did *not* consider man as a primary determinant—probably because he worked in an area in which nature was so overwhelmingly dominant. He portrayed towns, villages, fields and crops as elements of the landscape.

He envisaged and often described the landscape as a whole. He believed that it was only through the interconnections of phenomena that one could evaluate any one of them. His portrayal of Tenerife, Canary Islands and the *Llanos* of Venezuela have been quoted so often that they may be regarded as classics. He repeatedly compared areas with similar landscapes in different parts of the world. The tall grass of the *Llanos* he described as steppe, since it is a treeless grass-covered plain. The steppe is a vegetation type which occurs throughout the world, though there is considerable variation in the kind of vegetation, location, extent, and, indeed, significance to man. He located and compared with the *Llanos*, the pampas, the Old World deserts, the Russian steppes, and tropical savannas, the American prairies, the tundra, and the heathlands of north-western Europe. We must remember that in 1800 the continental interiors were only vaguely known, the great botanical explorations had scarcely begun, and there were virtually no accurate continental maps and emphatically no maps whatever of the distribution of the data that Humboldt was seeking to define and compare as to extent and location. All this had to be done mentally. The essential point is that he not only recognized unique landscapes but also realized that they have general relationships and common genetic causes with similar areas in other parts of the earth. This is the core of all geographic work.

Humboldt's methods of recording and describing in the field are important in themselves. He revealed most clearly his concept of the areal interconnections of phenomena in the landscape in his attempts to summarise the unprecedented observations of the phenomenal changes of land and life with altitude in intertropical America. In a matter of days he traversed from the equatorial forest to the tundra, the equivalent of some 5,000 miles from the equator to the Arctic circle. He noted, measured, and mapped changes in relief, plant life, crops, tree and snow lines, in a way that had never been done before. To generalize these features (particularly important since no maps were available) he used the cross section or profile. On a section line he showed the combined features and the way in which their nature and associations changed with increasing altitude. This appeared as the key to much of his description of Mexico, in which he recorded, for example, the distinction made in local usage between the *tierra caliente*, below 1,200 metres, with sugar, indigo, cotton, bananas, and the disease, yellow fever; the *tierra templada* (1,200–2,500 metres) with mid-latitude crops and the soil most suitable for human settlement, covering most of the high undulating plateaus; and the *tierra fria*, above about 2,000 metres. Humboldt then took the step of trying to portray the altitudinal changes he had observed. In Paris, in 1804, he had a Viennese landscape painter draw the cross section. In this, the *Naturgemälde der Tropenländer*, Humboldt sought to give pictorial representation to the generalized altitudinal zones between latitudes 10° north and south. He wrote as follows about the diagram[1]:

> I put together in this *Naturgemälde* all the phenomena that appear on the surface of our planet and the atmosphere that encloses it . . . The work aims at embracing all that changes with elevation . . . air temperatures, electrical charges, humidity, oxygen content, air colour (degree of blueness), geology, cultivation, animals . . . all are based on my own observations . . . The empiricist counts and measures what the phenomena directly reveal.

Here too we find repeated the objective of his field observations: 'My eyes shall be steadily fixed on the interaction of forces, the influence of inorganic creation on animate animal and plant life, upon this harmony.' This is what he meant by *physique du monde* of which he wrote as early as 1796.

When he travelled he had with him forty different kinds of instruments, including three-foot telescopes, sextants, quadrants, cyanometer (for measuring blueness of the sky) and barometers. He claimed that he could give the latitude and longitude of all the places at which he took measurements or collected specimens. These are the data that are assembled on the generalized diagram. His object was not simply to measure one kind of phenomenon, *but to bring out the*

[1] Alexander von Humboldt, *Ideen zu einer Geographie der Pflanzen nebst einem Naturgemälde der Tropenländer*, Tübingen 1807.

ways in which the great variety of observable phenomena of the landscape are associated and interconnected with each other at different places. This is the essence of his concept and the diagram noted above clearly shows its method and purpose in execution.[1]

Humboldt is generally and justly ranked as a founder of physical and plant geography, but he was above all a regionalist, in the sense that he recognized the interdependence of areal phenomena and the need for explaining any one set of spatially distributed phenomenon in relation to their spatial context. This runs through all his work but is particularly noteworthy in his classic essay on Mexico, wherein the focus of his investigation is on the condition and prospects of its people and the state. For the first time, he used population and other data for the constituent political divisions of the state in order to bring out similarities and differences between them. Thus, he indicated the twelve Intendancies and three other districts into which Mexico was divided. He recognized three divisions as in the temperate zone—the northern interior, the northwest oriented toward the Pacific, and the northeast oriented toward the Gulf of Mexico. The torrid zone embraced the central region and the areas of the southwest. If these same administrative units are considered according to their commercial relations—that is, their situation in relation to access to the coasts—they fall into three groups: the provinces of the interior, which do not extend to the oceans; the maritime provinces of the eastern coast; and the maritime provinces of the western coast. He wrote that 'these divisions, will one day possess great political interest, when the cultivation of Mexico shall be less concentrated on the central table-land or ridge of the Cordillera, and when the coasts shall become more populous'.

This work on Mexico is a landmark in geographic description. It utilizes census data by constituent administrative divisions. It emphasizes the contrasts of relief, climate, and production and shows the relation of these to the distribution of population (numbers, ethnic character, disease) and of cultivated products. The writer has a method and purpose that gives throughout a standard of relevance that is often lacking in the lengthy descriptions of Ritter. This study, the first of the systematic geographical descriptions, is in striking contrast to the encyclopedic compilations of writers from the days of Strabo to the nineteenth century topographers.

It is fitting to conclude with some comments on Humboldt's great single work, the *Cosmos*, for this was intended for wide consumption. The first two volumes were published in German in 1845

[1] R. Bitterling, 'Alexander von Humboldt's Amerikareise in zeitgenossischer Darstellung,' *Pet. Mitt.*, 98, (1954), pp. 161–71.

and 1847, the third and fourth in 1850 and 1858, and the fifth, unfinished, in 1862. The work has been translated into eight languages. As early as 1834 Humboldt outlined the work to a friend.

I have the crazy notion to depict the entire material universe, . . . from spiral nebulae to the geography of mosses and granite rocks, in one work—and in a vivid language that will stimulate and elicit feeling. . . . Fifteen years ago I had started writing it in French, and called it *Essai sur la physique du monde.* In Germany I had originally thought of calling it *The Book of Nature,* after the one that Albertus Magnus wrote in the Middle Ages. But all this is uncertain. Now my title is: *Cosmos, Sketch of a Physical Description of the Universe, after the enlarged lectures in the years 1827 to 1828.*

In the introduction of the *Cosmos* we read:

The most important aim of all physical science is this: to recognise unity in diversity, to comprehend all the single aspects as revealed by the discoveries of the last epochs, to judge single phenomena separately without surrendering to their bulk, and to grasp nature's essence under the cover of outward appearances. . . . The purpose of this introductory chapter is to indicate the manner in which natural science can be endowed with a higher purpose *through which all phenomena and energies are revealed as one entity* pulsating with inner life. Nature is not dead matter. She is, as Schelling expressed it, the sacred and primary force.

The philosophy of the *Cosmos* comprises four main themes. The first is the definition and limitation of a physical description of the world as a special and separate discipline. The second is the objective content, which is the actual and empirical aspects of nature's entity in the scientific form of a portrait of nature. Third, the action of nature on the imagination and emotion becomes an incentive to nature studies through media such as travel descriptions, poetry, landscape painting, and the display of contrasting groups of exotic plants. And last, the history of natural philosophy and the gradual emergence of concepts pertaining to the cosmos as an organic unit are dealt with.

Humboldt's *Cosmos* is a monument of compilation and in it are embodied the conclusions drawn from a lifetime of travel and diligent research. The method of presentation consists of 'the art of collecting and arranging a mass of isolated facts and rising thence by a process of induction to general ideas'. Humboldt was unable to incorporate new developments in science owing to the long period which elapsed between the publication of the first and last volumes. For instance Helmholtz's conservation of force, Joule's mechanical theory of heat, and the spectrum analysis developed in the eighteen sixties were all ignored, or partially embodied, or criticized in the work, and in certain respects these works blighted some of Humboldt's theories.

Several concluding comments should be made about the work of this remarkable man. First, Humboldt regarded man as a constituent part of the universe. Man formed one item in the balanced unity of nature which his capacities allowed him to contemplate with aesthetic satisfaction. Hence Humboldt's emphasis in *Cosmos* on the history of art as man's interpretation of nature. This experience embodies 'the principle of individual and political freedom forever rooted in the equal rights of a unified humanity'.

A second point concerns Humboldt's view of the earth itself and man's place in it.

> The general view of nature which I have endeavoured to present would be incomplete, were I to chose it without attempting to trace, by a few characteristic traits, a corresponding sketch of *man*, viewed in respect to physical gradations, to the geographical distribution of contemporaneous types, to the influences which terrestrial forces exercise on him, and to the reciprocal but less powerful action which he in turn exerts on them. Subject, though in a lesser degree than plants and animals, to the circumstances of the soil and the meteorological conditions of the atmosphere, and escaping from the control of natural influences by activity of mind and the progressive advance of intelligence, as well as by a marvellous flexibility of organisation which adapts itself to every climate, man forms everywhere an essential portion of the life which animates the globe.[1]

Third, Humboldt, like Ritter, believed in the concept of the unity of nature including man. While Ritter's concept was a teleological one, in which the earth was created as part of a grand design so as to fit the needs of man, Humboldt believed in the balance of nature of which man was an integral part. Both believed, like the generation of romanticists to which they belonged (including, for example, Goethe, a contemporary and friend of Humboldt) in the organic coherence of all phenomena and in the earth as an integrated organic whole. This unity had a divine purpose for Ritter, with its end in the welfare of man, but Humboldt considered man objectively as an

[1] Alexander von Humboldt, *Cosmos: Sketch of a Physical Description of the Universe.* Translated under the superintendence of Edward Sabine, 4 vols. London. 1846–58. Vol. 1 p. 350.

Alexander von Humboldt

element in the balance of the physical world and interpreted the unity in terms of aesthetic interpretation and values.

The great French geographer, Emanuel De Martonne, writes as follows in his classic work on physical geography, first published in 1909.[1]

> Whatever phenomena he studied, relief, temperature, vegetation, Humboldt did not merely treat each individually as a geologist, meteorologist, or botanist. His philosophical outlook carried him further. It led at once to the observation of other phenomena. He sought causes and distant consequences, even including political and historical facts. Nobody has shown with more precision how man depends on the soil, climate, vegetation, and how vegetation is a function of physical phenomena, and how they all depend on each other.

The same writer claimed[2] that Humboldt was the first scholar clearly to define and apply what he (De Martonne) considers to be the two essential principles that make geography 'an original (distinctive) science', rather than a collection of facts from the physical and biological sciences. Humboldt, he writes, sought not the individual spatial phenomena, but the complex of spatially arranged phenomena. Humboldt observes these phenomena and then seeks to explain them: '*il remonte vers les causes et redescend jusqu'au conséquences . . .*'. This is described by De Martonne as the principle of causality. A second principle he defines as the principle of general geography, by which he means that Humboldt sought to compare the location and extent of terrestrial phenomena on the face of the earth.

Specialists have recognised Humboldt's important contributions to their particular fields. What has not been so generally recognised among English speaking scholars, is Humboldt's insistence on the areal associations of diverse categories of physical and human phenomena. This is his contribution to geographic knowledge and the discipline of geography. In this respect he was a pioneer and well ahead of his times and resources.

Further References

A standard work, though old, on the life of Humboldt is Karl Bruhn's *Alexander von Humboldt, Eine wissenschaftliche Biographie*,

[1] E. de Martonne, *Traité de Géographie Physique*. Tome I, 1948.
[2] *Ibid*, pp. 15–16.

1872, three volumes. I have also drawn freely from the following sources—Döring, *Wesen und Aufgaben der Geographie bei Alexander von Humboldt, Frankfurter Geographische Hefte*, 1951 (with many quotations from Humboldt's works and a bibliography); Helmut de Terra, *The Life and Times of Alexander von Humboldt*, New York, 1955; Carl Troll, 'The work of Alexander von Humboldt and Carl Ritter: A centenary address', *Advancement of Science*, 16, No. 64, (1960) pp. 441–52; and Karl A. Sinnhuber, 'Alexander von Humboldt, 1769–1859', *Scottish Geog. Magazine*, 75, (1953) pp. 83–101. The memorial works published for the centenary of Humboldt are J. H. Schultze (ed.), *Alexander von Humboldt: Studien zu seiner universellen Geisteshaltung*, Berlin, 1959 (West German memorial volume); Kommission der Deutschen Akademie der Wissenschaften (ed.), *Alexander von Humboldt, 14. 9. 1769–6. 5. 1859*, Berlin, 1959 (East German memorial volume). A definitive work is H. Beck's *Alexander von Humboldt*, two volumes, Wiesbaden, 1959 and 1961, an authoritative biography with emphasis on the geographical significance of Humboldt's work. (*Von der Bildungsreise zur Forschungsreise*, 1769–1894 (1959) and *Vom Reisewerk zum Kosmos*, 1804–59 (1961)). Two lesser works are L. Kellner, *Alexander von Humboldt*, London, 1963, and W. T. Stearn, 'Alexander von Humboldt and plant geography,' *New Scientist*, 5, (1959) No. 128, pp. 957–9. For an exhaustive bibliography of the works of Humboldt see J. Lowenberg, *Alexander von Humboldt, Bibliographische Ubersicht seiner Werke*, Stuttgart, 1960. This contains 639 entries.

Works by Humboldt himself obtainable in English translations are the following as discussed in the text:

Cosmos: Sketch of a Physical description of the Universe, translated by E. Sabine, 4 vols., London, 1846–58.

Cosmos, translated by E. C. Otté and others, London, 1849–58, 2nd edition, 1871–83.

Personal narrative of travels to the equinoctial regions of the New Continent during the years 1799–1804, translated by H. M. Williams, 7 vols., London, 1818–23.

Personal narrative of travels etc., translated by T. Rosa, 3 vols., London, 1852–69.

Views of Nature, translated by E. C. Otté and H. G. Bohn, London, 1847.

Aspects of Nature, translated by E. Sabine, 2 vols., London, 1849.

A geognostical essay on the superposition of rocks in both hemispheres, translated from French, London, 1823.

Political essay on the kingdom of New Spain, translated by J. Black, 4 vols., London, 1811–22.

Alexander von Humboldt

Researches concerning the institutions and monuments of the ancient inhabitants of America, translated by H. M. Williams, 2 vols., London, 1814.

Special attention is drawn to two recent articles in German periodicals by senior geographers: O. Schmieder 'Alexander von Humboldt, Persönlickheit, wissenshaftliches Werk and Auswirkung auf die moderne Länderkunde', *Geographische Zeitschrift*, 52 (1964), pp. 81–95; and G. Pfeifer, 'Alexander von Humboldt (1859–1959): Beiträge zur Würdigung seiner Persönlichkeit anlässlich der Gedenkfeiern in Süd und Mittelamerika im Jahre 1959', *Sitzungsberichte der Physikalische—Medizinischen Sozietät zu Erlangen*, 80 (1959).

3

Carl Ritter (1779-1859)

Carl Ritter[1] exercised a much more direct influence on the growth of geography in Germany than did Humboldt. Born in Quedlinburg in 1779, he began his academic studies at the University of Hallé (a proud claim of that University to this day). In October 1798, he became the private tutor to the two sons of a wealthy banker in Frankfurt, and in an atmosphere of luxury and leisure he met many interesting people and was able to pursue his studies in comfort. An important change in his attitude to human history was brought about by his first meeting in 1807, with the great Swiss educator, H. Pestalozzi, and thereafter he turned to matters of 'general geography'. He accompanied one of his protégés to the University of Göttingen in 1813 and here he continued his own research. This concerned the first volume of the *Erdkunde* which was published in 1817, and which, curiously enough, handled one of the least known portions of the earth, the dark continent of Africa. On the strength of the stir caused by this work he was offered a professorship in geography at the University and at the Royal Military Academy in Berlin, a post which he held until his death in 1859. During this long period of almost forty years he had a profound influence on the thinking of people in many walks of life, and not least on those in high places in the army. He did not resume the writing of his *Erdkunde* until 1832, and at the time of his death this had reached 19 volumes, most of which were concerned with Asia. He had not even started to deal with Europe, the best known of all the areas of the earth.

THE WORKS OF RITTER

Ritter's first geographical works were two volumes on Europe which appeared in 1804 and 1807. These were supplemented in 1806 by a

[1] Karl A. Sinnhuber, 'Carl Ritter, 1779–1859', *Scottish Geographical Magazine*, 75 (1959), pp. 152–63.

series of six maps of the continent. The text was in the traditional arrangement, but the atlas was a real contribution which showed on simple maps the mountain ranges, the vegetation (trees and shrubs), the cultivated plants in relation to climate, the distribution of wild and domesticated animals and nationalities.

In 1810 Ritter prepared a manuscript on what he called the new geography. Though it was never published its ideas underlie much of his later work. The title is *Handbuch der allgemeinen Erdkunde oder die Erde, ein Beitrag zur Begründung der Geographie als Wissenschaft*. Sinnhuber[1] has summarised this work as follows:

> The study of geography is of educational value not for its usefulness but because it studies a phenomenon of nature, the earth's surface. This as a whole, as in its parts, shows an inherent plan, thus every detail is of importance. To find out its design will take infinite painstaking research and success will only come if the study is carried out without bias and no *a priori* systems are introduced from without.

This expresses clearly the inductive method of Ritter and his rejection of *a priori* theory. He considered geography to be an empirical and descriptive science.

Geography, he wrote, deals with local conditions (*Lokalverhaltnisse*) and embraces the attributes of place with respect to topical, formal, and material characteristics. The first attribute was topographical, that is, it dealt with the natural divisions of the earth's surface and was intended for teaching presentation according to the principles of Pestalozzi. The second included the distribution and movements of water, sea, and atmosphere; the bases of human life. The material conditions were described as the geographical aspect of natural history; this covered the distribution of minerals, plants, and animals.[2]

It was during his two years sojourn at Göttingen that Ritter began to write the first volume of the *Erdkunde*. His approach, he said, had entirely changed, and he claimed that its distinctive theme was a comparative one which aimed at showing the connection between history and nature, both organic and inorganic. He wrote: 'I have demonstrated that geography has a right to be considered a sharply-defined science, of kindred dignity with the others.'[3] He was engaged on the first volume for eight years. It was published in 1817 and

[1] *Ibid.*

[2] H. Schmitthenner, Carl Ritter, *Frankfurter Geog. Hefte*, Heft 4, 1951, p. 48 *et infra*, especially pp. 51–2.

[3] Quoted in W. L. Gage, *Life of Ritter*, Edinburgh, 1867, p. 141.

D

dedicated to Pestalozzi and his early tutor, J. C. F. Gutsmuths. It should be noted that Ritter's next work, written at Frankfurt, was essentially a historical tract dealing with the shifts of peoples before the time of Herodotus in the area of the Caucasus and the Black Sea. It was intended to be a link between the volume of the *Erdkunde* on Asia and the next volume on Europe, though, as noted above, he never reached the treatment of Europe.

A revised edition of the first volume of the *Erdkunde* (on Africa) appeared in 1822. From 1832 on he published new volumes regularly until his death and the work ran into nineteen volumes and 20,000 pages, but as a world survey, it was incomplete. The full title runs as follows in German: *Die Erdkunde in Verhältnis zur Natur und zur Geschichte des Menschen oder allgemeine vergleichende Geographie als sichere Grundlage des Studiums und Unterrichts in phyikalischen und historischen Wissenschaften*. The English translation would read as follows: *Earth Science in relation to Nature and to the history of Man: A general comparative geography, as a secure foundation for study and teaching in the physical and historical sciences*. The fourteen volumes on Palestine and the Sinai peninsula were translated and condensed by William L. Gage with the title *The Comparative Geography of Palestine and the Sinaitic Peninsula* (4 vols., Edinburgh, 1866).

The meaning of the term 'comparative geography' has been much disputed, since Ritter himself was vague about it.[1]

Ritter[2] himself claimed that:

The very word Geography, meaning a description of the Earth, has unfortunately been at fault, and has misled the world: to us it merely hints at the elements, the factors of what is the true Science (*Wissenschaft*) of Geography. That science aims at nothing less than to embrace the most complete and the most cosmical view of the Earth; to sum up and organize into a beautiful unity *all* that we know of the globe. . . . Geography is the department of science that deals with the globe in all its features, phenomena, and relations, as an independent unit, and shows the connection of this unified whole with man and with man's Creator.

He claimed that the central principle of geography is 'the relation of all the phenomena and forms of nature to the human race', examined and organized within the framework of the unique geographical associations of land and man on the earth's surface. To

[1] Sinnhuber, *op. cit.*
[2] Carl Ritter, *Comparative Geography*, translated by W. L. Gage, Cincinnati and New York, 1864, pp. 19–20 London, Edinburgh, 1865, pp. 16–17.

this end, geography must draw upon all the kindred sciences of earth and man. There follows a statement that should figure in the mind of every student of geography.

> It is to use the whole circle of sciences to illustrate its own individuality (that is the uniqueness of terrestrial areas in terms of land-man relations) not to exhibit their peculiarities. It must make them all give a portion, not the whole, and yet must keep itself single and clear.[1]

Though soaring here into the realms of teleological interpretation, Ritter also clearly envisaged, as a craftsman, the work of the geographer. He recognized the peculiar place of geography in the realm of science, drawing upon relevant data in order to achieve its own unique objectives.

Ritter makes a rhetorical claim for geography as 'the science of the earth' (in the sense of natural philosophy or cosmology) that reaches far beyond his real objectives, namely, the description of the earth as the home of man. This approach, as a philosophy and a craft, was the constant purpose of the *Erdkunde*, in which is locked most of the work of Ritter. He asserted that the *Erdkunde* would definitely establish geography as an organized body of knowledge. The superiority of his work over that of his predecessors was expressed as follows:[2]

> My aim has not been merely to collect and arrange a larger mass of materials than any predecessor, but to trace the *general laws* which underlie all the diversity of nature, to show their connection with every fact taken singly, and to indicate on a purely historical field the perfect unity and harmony which exist in the apparent diversity and caprice which prevail on the globe, and which seem most marked in the mutual relations of nature and man. Out of this course of study there springs the science of physical geography, in which are to be traced all the laws and conditions under whose influence the great diversity in things, nations, and individuals, first springs into existence, and undergoes all its subsequent modifications.

Ritter believed the earth to be an organism (*organische Einheit*), made, even in its smallest details, with divine intent, to fit the needs of man to perfection: 'as the body is made for the soul, so is the physical globe made for mankind'. He envisaged the time when it would be possible 'for men sending their glance backwards and

---

[1] *Ibid*, p. 27.
[2] Letter quoted by Gage, *Life of Ritter*, p. 144.

forwards, to determine from the whole of a nation's surroundings, what the course of its development is to be and to indicate, in advance of history, what ways it must take to retain the welfare which Providence has appointed for every nation whose direction is right and whose conformity to law is constant.'[1]

Such vague expressions of a teleological philosophy should not divert our attention from Ritter's professional work. A lifetime of reading, selection, interpretation, and organization of a mass of available data was spent in writing the *Erdkunde*. The above philosophical speculations were contained in his general lectures and in his correspondence. For his working concepts we must turn to his greatest substantive works, the *Erdkunde*.

I make no claim to have read the nineteen volumes of the *Erdkunde* in their entirety. In many ways, says Schmitthenner, it can be compared, as a standard work to Eduard Suess' later geological work on the face of the earth. The work is vast in its content and often vague as to procedure. For this reason I rely on German geographers who can claim to speak with authorative judgement. In this respect I have found particularly helpful the monograph by a senior and highly respected German geographer, Heinrich Schmitthenner. In addition, the memorial volume, published on the occasion of Ritter's centenary, is indispensable, especially the essay by Ernst Plewe.

Ritter recognized the validity of Humboldt's idea of a terrestrial unit as an organic whole (*organische Einheit*), as exemplified in the latter's description of the *Llanos* and the steppes. He sought to develop the concept of terrestrial or spatial unity (*Raumbegriff*). Two comments alone will be made here. Ritter, mainly on traditional and also on religious grounds, recognized the division of the earth organism into major continental units or *Erdteile*—both natural units (e.g. Europe) and man-named units (e.g. the Old and New World). Each of these falls into major physical (natural) divisions. But the smallest units need to be analysed as to the *dingliche Erfüllung* of the earth's space so as to arrive at generalizations as to the earth's mosaic of terrestrial spaces by inductive reasoning. The second point is that this kind of study of the earth's association of distinct features is essentially geographic. It is *Länderkunde*. Generalizations of world wide significance (according to Ritter) would seek to typify the areal expression of the factors—natural and human—that account for the differentiations of the earth's surface, geofactors, as Schmitthenner described them. This is the field of *Allgemeine Erdkunde* and at the time that Ritter lived there were neither the data nor the maps avail-

[1] Carl Ritter, *Comparative Geography*.

able to permit but the most tentative generalizations of this order. This involved Ritter in teleological speculation, but such study must advance, he always argued, by collection and generalization of data about particular localities. Yet he ultimately aspired to the development of fields of world comparison, for which he was criticized. He wrote on the field of general geography, and, in his general lectures, gave clear indications as to how *Allgemeine Erdkunde* should be developed—physical geography, the areal distribution and impact of economic production, and the spread and cultures of human societies. He envisaged the development of the 'mobile forms' (in his second group of data) of oceanography, climatology, vulcanicity, and plant and animal geography (Schmitthenner, *op. cit.*, p. 66). These mobile and fixed forms were the elements of the 'total environment' in which man and his works are to be evaluated.

It is important to emphasize that such general study, or, as we would say today, systematic study, seeks to determine the exact distribution and types of areas in which the same geofactors are operative in the characterization of distinctive areal associations of phenomena. The way ahead was indicated clearly in several remarkable essays by Ritter, that are available in English, unreliable as the translations by Gage sometimes are, in the *Comparative Geography* and *Geographical Studies*. His essays on the 'fixed forms' of the earth's surface, on the historical element in geographical science (1833) and on the resources of the earth (1836) are among the most significant and foresighted of his writings that are still eminently valuable to understanding of the geographic viewpoint.

The general content and plan of the *Erdkunde* is discussed by Schmitthenner in Chapter 5 of his *Studien über Carl Ritter* (Ch. 5, p. 63 *et infra*). He says that Ritter had already made this clear in his writings of 1806 and 1809–10. The *Erdkunde* was to embrace three major topics—topical (*topischen*), formal (*formellen*), and material (*materiellen*). The first deals with the fixed forms of the continents, that is with the distinct divisions of the earth's surface. The second set of terrestrial facts embraces the 'mobile forms' of each *Erdteile* on the basis of essential elements of water, air and fire. The third set of facts embraces the three realms of Nature, only in so far as they have localized distributions that are relevant to other phenomena on the earth's surface. They are interpreted in terms of type areas, in relation to natural physical areas; their localization and spread over the earth; the degree of dominance in areas of terrestrial impact in the world the result of forces of Nature or Man. In the second and third aspects Ritter sought to develop oceanography, climatology, vulcanicity, plant and animal geography as a first and essential step to his geographic goal. Each needs to be interpreted in areal distribution in relation to other spatially distributed phenomena so as to throw light on the distinctive associations of phenomena in space.

The regional concept as developed by Ritter is the matter of main interest to geographers today. The major geographic units are the continents or *Erdteile*. Here he simply accepted traditional definitions. Each continent was subdivided by its orography into a highland core, surrounded by terrace lands, drained by the major rivers, and peripheral coastal lowlands. He apparently believed that each continent had a similar build. Therefore, these are divisions of a second order arrived at deductively. Lesser units, which might be called third order, are arrived at from the detailed configuration and content (*dingliche Erfüllung*) of particular areas. They are the bricks or units of the terrestrial mosaic upon which generalizations must be based and are thus arrived at inductively. Thus, Ritter appears to have envisaged each continent as a complex of individual units (*Individuen*) whose whole structure and extent could be arrived at by inductive procedures from the smallest areas of associated phenomena (*Individualitäten*). He never used the term *Landschaft*. Heinrich Schmitthenner writes that Ritter seeks out 'the areal localities and orients himself as exactly as possible in them. These localities he groups together according to the individual phenomena and the forces that operate in them simultaneously. Then with the complex of various groups, he aims at general concepts in relation to physical and other factors of each locale to the organic.'[1]

Ritter based his whole system on Africa and Asia, the first two continents he studied. It was on the basis of their configuration that he arrived at a classification of landforms, and conceived each continent as being built on a similar plan. First are the highlands and plateaus, which are of two orders, namely those regions with an average elevation of 4,000 to 5,000 feet and those with a lower average elevation. Second are the mountains, which are divided into five groups: (a) parallel mountain chains, such as the Jura and Himalayas; (b) diverging or converging mountain chains, e.g. the eastern Alps and northern Rockies (divergence) and the mountain knots of the Andes (convergence); (c) ranges radiating from a central nucleus, e.g. Auvergne southwest Alps; (d) ring-shaped systems, e.g. Transylvania and Bohemia; (e) cross mountains, in which two or more ranges meet at high angles, e.g. the Hindu Kush and Himalayas, the Kwen Lun and the Pamirs. Third are the lowlands, with an elevation below 400 feet, an example of which is the great European plain. Fourth are the regions of transition between

[1] H. Schmitthenner, *op. cit.*, Ch. 5, p. 58.

highlands and lowlands, called lands of gradation or terrace lands.

This classification, summed up in the introduction to the *Erdkunde*, formed the basis of Ritter's treatment of the continents. Superficial as the scheme may be, it afforded a new regional method of description which was distinct from the usual method of dealing with political units. In each division the main features of the relief and drainage are described, followed by climate, major products, and population. Some historical events and records of journeys of exploration are included. It concludes with a general summary.

It is appropriate to comment on both the title and subtitle of the *Erdkunde*, for herein lies much of the conceptual difficulty in the development of geography in the second half of the century. First, geography is not the equivalent of *Erdkunde* as the *science of the earth*. Earth science involves the shape, size, and structure of the earth, and these subjects fall to geology and geophysics. In Ritter's view geography is concerned with the earth's surface in terms of the areal differences of spatially associated phenomena upon it. The study of the world-wide distribution of a single phenomenon may well fall into the realm of an ancillary discipline with a spatial aspect. The uniqueness of geography lies in characterizing and explaining the areal associations of terrestrial phenomena in unit areas of different orders beginning with the smallest areas and ending with major generic regions.

The second point concerns the subtitle—*im Verhältnis zur Natur und zur Geschichte des Menschen, oder allgemeine vergleichende Geographie*. This raises a question which geographers since the eighteenth century have sought to resolve. It is of great importance since it is the key to the structure of the discipline and its relation to the natural and social sciences. If the earth is studied as the home of man, then land, water, atmosphere, and the world of plants and animals are studied as the human habitat, strictly with reference to human occupance. If, on the other hand, all the realms are studied as ends in themselves, then landforms, climatology, oceanography, and biogeography must be studied irrespective of human occupance. It is true that the more one knows of physical processes the better one can evaluate the significance of physical features to human occupance, but this is not the question at issue. The anthropocentric view requires data about the physical earth that are relevant to man. This means questions of soil types, incidence of erosion and conservation problems, depth of water table and drainage, and so forth. Study of the physical earth, irrespective of Man, which was apparently embraced in the conceptual framework of Ritter, today involves, with the great advance of contemporary knowledge, independent study of the origins and development of landforms, and plant and animal ecology, each of which is far removed from study of the physical earth and its relevance to human societies.

Ritter has been accused of making geography 'a handmaiden to history' and there were those who, in the decades after his death, wished to throw man out of geography altogether and confine the study to the physical aspects of land, water, and atmosphere. Subsequently a compromise was found in the regional concept that took shape among professional geographers in the turn of the century.

CONCEPTS

Let us now summarize Ritter's main geographical concepts.

First, Ritter conceived of geography as an empirical science rather than one based on deduction from rational principles or from *a priori* theories. He saw geographic study as proceeding from observation to observation, and although he was convinced that there were laws, he was in no hurry to establish them, as the enormous compilation of the *Erdkunde* shows.[1]

Secondly, there is a coherence in the spatial arrangement of terrestrial phenomena which is described by both Ritter and Humboldt as *Zusammenhang*. Areal phenomena are so interrelated as to give rise to the uniqueness of areas as individual units. Ritter, therefore, considered that areal synthesis and description, must precede the world wide analysis of particular sets of phenomena.

'The earth and its inhabitants stand in the closest mutual relations, and one element cannot be seen in all its phases without the others. On this account history and geography must always go hand in hand. The country works upon the people, and the people upon the country.'[2]

Geography must rise above mere description. To avoid the 'diversity of origin of its materials' it must have a unique and guiding central principle. 'It is no use the whole circle of sciences to illustrate its own individuality, not to exhibit their peculiarities. It must make them all give a portion, not the whole, and yet must keep itself single and clear.'[3] He sought for *relations* or *connections* between sets of phenomena in the same area and between one place and another. The task of geography is 'to get away from mere description to the law of the thing described; to reach not a mere

[1] R. Hartshorne, *op. cit.*, p. 230.
[2] Preface to 'Europe', 1804, quoted by Gage, *op. cit.*, p. 113.
[3] Carl Ritter, *Comparative Geography*, p. 28.

enumeration of facts and figures, but the connection of place with place, and the laws which bind together local and general phenomena of the earth's surface' and again 'what then is the task imposed upon geography, in its work of analysis, but to reach the connection which exists between parts; or, in other words, to get at the relation between places and what fills and occupies them'[1]. The geographer traces '*causation* and *interdependence* of the (spatially distributed) phenomena, and the *relations* of every one to the country which supplies its conditions of being'.[2] These words give the keynote to Ritter's approach.

Thirdly, Ritter insisted that boundary lines whether wet or dry such as rivers or mountains, were 'a means towards the real purpose of geography, which is the understanding of the content of areas'.[3] Geography, he maintained, is the study of *der irdisch erfüllten Raüme der Erdoberfläche*.

Fourthly, geography according to both Humboldt and Ritter, was concerned with objects on the earth as they exist together in area. Hartshorne writes that Ritter considered 'that as chronology provides the framework into which the multiplicity of historical facts are ordered, the area (*Raum*) provides the skeleton for geography; both fields are concerned with integrating different kinds of phenomena together, each in its respective frame'. To pursue this end, Humboldt studied systematically particular sets of phenomena in their areal relations with other phenomena. Ritter studied areas synthetically, that is, in their totality. He deliberately deferred the formulation of laws till later.[4]

Fifthly, Ritter held a holistic view with respect to the content and purpose of geographic study and the whole study was focussed on and culminated in man. He wrote that the purpose of the *Erdkunde* was 'to present the generally most important geographic-physical conditions of the earth-surface in their natural coherent interrelation (*Naturzusammenhang*), and that (the earth-surface) in terms of its most essential characters and main outlines, especially as the fatherland of the peoples in its most manifold influence on humanity developing in body and mind.'[5] But this, Ritter claimed in the introduction to the *Erdkunde*, was not an exclusive concern with man. 'Independent of man, the earth is also without him, and before him, the scene of the natural phenomena; the law of its formation cannot

[1] See essay on the 'Historical Element in Geographical Science, in *Geographical Studies*, trans. by W. L. Gage, New York 1861, p. 246 and p. 247.

[2] *Ibid*, p. 22.

[3] R. Hartshorne, *The Nature of Geography*, p. 233.

[4] *Ibid*, p. 233.

[5] *Ibid*, p. 238.

proceed from man. In a science of the earth, the earth itself must be asked for its laws.'[1] The dilemma presented by this double viewpoint has already been noticed.

Sixthly, we turn to Ritter's concept of the unique and distinct geographical unit (*Individuen*). In his earlier years, Ritter, following Buache and Gatterer, adhered to the idea of the river basin and its bounding mountain ranges as natural units. It was later that he made the discovery of such units as the objective of study. In 1806 he expressed the notion of the organic unity of geographic areas. 'Every naturally bounded area is a unity in respect of climate, production, culture, population and history.'[2] This deterministic view was changed later by the idea that the degree of cohesion of these phenomena is the objective of study. He came to regard the geographical unit not as something given by a natural framework, but as something to be discovered in the physiognomy of the earth.[3]

The study of a geographic unit as a spatial concept was an essential purpose of Ritter's geography and is best illustrated in his volume on Africa. He recognised major physical divisions of each continent on a deductive basis. The earth he regarded as an organic entity (*organische Einheit*). In the further breakdown he arrived at distinct units (*Erdteile*) according to the major features of physical demarcation. The continent or *Erdteil* is not a generic complex. It is a complex of individual units (*Ländersysteme*). The geographer, he wrote, seeks to examine every earth area according to its characteristics, without approaching these on a one-sided logical system of division and classification. He is concerned with the total content of areas. This content, viewed geographically, embraced what he called the fixed forms, the mobile forms (atmosphere, hydrosphere, vulcanicity), and the material forms. This makes clear the concept of areal study (*räumliche Prinzip*) as opposed to systematic study (*sachlich systematisch*). In fact, he rejects the latter as not geographic (and this in spite of his frequent declaration of general laws of land-man relations). Ritter sought to build up systems of areas inductively. The ways in which these various sets of phenomena (for example, plant formations and climate) are spatially interdigitated and interdependent, he was never really clear about.[4]

Finally, Ritter's philosophical viewpoint was teleological. Although he insisted that geography investigates the spatial variations of the total content of terrestrial areas, he also believed that the earth

---

[1] *Ibid*, p. 239.
[2] A. Schmitthenner, *op. cit.*, p. 53.
[3] *Ibid*, p. 53. The whole of section 5 is concerned with this theme.
[4] *Ibid*, p. 57.

was designed to serve one end—namely, as the abode of man. This viewpoint, expressed repeatedly in his writings, was based on his religious convictions, as well as on his primary interest in the history of man. It was his philosophic interpretation of what he could not understand. It had no effect on the method and substance of his work as contained in his masterpiece of geographic compilation, the *Erdkunde*,[1] but this was the approach that dominated the geographic viewpoint of his successors, and brought the study into disrepute.

In conclusion, we quote the following from an obituary address by the President of the Royal Geographical Society in 1860.

> The labours of Carl Ritter are characterised by great industry, and an anxious desire to gather up, and systematically to arrange, every fact relating to the regions treated of in his work, and to leave no source unexplored from which any information was to be derived. His great work comprises not only the geography of each country strictly considered, but also the history, antiquities, politics, ethnology, natural history, and an account of any travels through them which may tend to throw light upon their condition.[2]

This is intended as a tribute to Ritter's scholarship, but it emphatically does not reveal a real understanding of the geographic concepts which Ritter sought to develop. It emphasises Ritter's eclectic approach and his unduly strong emphasis on history. This was a weakness as we shall see later in the contemporary reactions to it and the scientific developments in the geography of lands and peoples in the next generation.

The main contribution of Ritter to human knowledge was his concept of the regional associations of terrestrial phenomena at various levels over the earth's surface. He never recognized any basic distinction between *Länderkunde*—the geographic study of particular areas—and *Allgemeine Erdkunde*. They were different sides of the same coin, general geography dealing with the character, typology, location and extent of different categories of terrestrial phenomena, natural (including wild plants and animals) and man. There are here two different concepts, the one chorological, the other ecological with respect to Man. The advancement of knowledge since Ritter's day has heightened this difference, and has been, and still is, the basis of much controversy.

[1] A. Schmitthenner, *ibid.*, Ch. 6 and R. Hartshorne, *op. cit.*, p. 3.
[2] Earl de Grey and Ripon, 'Carl Ritter', *Proceedings of the Royal Geographical Society*, 4 (1860), being an obituary address by the President of the R.G.S. on May 28th, 1860.

In conclusion, two basic concepts run through all Ritter's writings. The first is that geographical study seeks to establish and explain the areal variations of the earth's surface, in respect to Nature—land, water, air, vegetation and animal life—and in respect to Man by his presence and works. This approach focusses its attention on the differential character of the material content of terrestrial areas. It does not examine one category of facts in isolation, but seeks to find out and explain its association with other spatial distributions in order to arrive at the portrayal and explanation of areal complexes. The second concept regards Man as the focus of attention. This is an approach and a philosophical view point that preoccupied many philosophers and historians right through the nineteenth century. These two approaches imply in the first case, the study of the areal variables of Man and his works on a footing of equality with the other realms of Nature; and, in the second case, the evaluation of the spatial complexes presented by Nature in their relevance to Man. These two approaches, the one essentially chorological, the other ecological, have been contended continuously ever since. Indeed, the difference between the two is no mere theoretical matter. It is a source of confusion to both some geographers and to natural and social scientists, who seek to know—and often jealously and ably dispute— what geographers are doing and where they stake out their claims within which they have special competence.

There is a second kind of dualism that remains a matter of professional concern and that comes out continuously in subsequent research activities and teaching. This is the relation between the study of particular areas, *Länderkunde*, and general world geography, *Allgemeine Geographie* or, as Ritter usually called it, *Erdkunde*. The latter is concerned with the world-wide distribution of particular categories of data on the earth's surface in respect to both facts and factors of areal differentiation. Ritter sought to develop both, regional and systematic, and while it is true that his *Erdkunde* was essentially concerned with particular areas, he made innumerable statements urging the development of a comparative world-wide approach of areas in his teaching and writings. Reference to Hartshorne's pronouncements alone demonstrates this; but one clearly finds it in the Gage translations of his addresses and writings to say nothing of his writings in German. He exposed the conceptual framework of both regional and world-wide geography. He lacked the data in his time to reach world-wide generalizations about areas, though he urged that they should be the geographer's ultimate goal. There was lacking in his day the knowledge of vast areas of the earth, data of land, air and water, and therefore of reliable maps, to permit typological generalizations. But he always encouraged the advancement of such an approach and its incorporation into the framework of the geographer.

Further References

This brief appraisal is based on recent thorough studies in English as follows: R. Hartshorne, *Nature of Geography*, pp. 48–84; Karl A. Sinnhuber, 'Carl Ritter, 1779–1859', *Scottish Geographical Magazine*, 75 (1959), pp. 152–63; and Carl Troll, 'The Work of Alexander von Humboldt and Carl Ritter', *Advancement of Science*, 16, No. 64, (1960), pp. 441–52. There is also much of value in G. Tatham, 'Geography in the Nineteenth Century', in G. Taylor (ed.), *Geography in the Twentieth Century*, London, 1951. Reference should also be made to J. Leighly's 'Methodological controversy in nineteenth century German geography', *Annals Assoc. Am. Georg.*, 28 (1938), pp. 238–53 and Carl Sauer's essay on Ritter in the *Encyclopedia of the Social Sciences*, New York, 1934, Vol. 13, p. 395.

Among recent works by German geographers we have drawn heavily on Heinrich Schmitthenner's *Studien über Carl Ritter*, *Frankfurter Geographische Hefte*, Heft 4, (1951), and E. Plewe and J. H. Schultze, 'Carl Ritter zum Gedächtnis', *Die Erde*, 90 (1959), Part 2. This is a special issue of the journal in commemoration of the centenary of Ritter's death, containing nine papers in German with English summaries. Also H. Beck, 'Carl Ritter Forschungen', *Erdkunde*, (1956), pp. 227–33 with extensive references. A bibliography of the writings of Carl Ritter will be found in *Die Erde*, 94 (1963), pp. 13–36.

English translations of Ritter's works are available as follows: The book entitled *Comparative Geography*, translated by W. L. Gage, New York, 1864, London, Edinburgh, 1865, contains lectures on General Geography as given in the University of Berlin and published as *Allgemeine Erdkunde* in 1862. A second book entitled *Geographical Studies* also translated by W. L. Gage, contains a series of public lectures, some of them delivered to the Berlin Geographical Society of which he was the founder and for many years its President. This book also contains the introduction to the first volume of the *Erdkunde*. Published first in Berlin in 1852, Gage's translation was published in New York in 1861. The original work is entitled *Einleitung zur allgemeinen vergleichenden Geographie*, Berlin, 1852. It contains important statements on 'the historical element in

47

geographical science' (1833) and 'remarks on the resources of the earth' (1836). The Gage translations, as Hartshorne rightly points out (*Perspective*, references 22 and 23), are 'not reliable', as the translations are literal and often do not convey the proper meaning of the author. Less well known are the following—*The Colonisation of New Zealand*, London, 1842 (translation of a paper read for the Wissenschaftliche Verein zu Berlin on 22nd Jan., 1842); *The Comparative Geography of Palestine and the Sinaitic Peninsula*, translated and adapted for the use of biblical studies by W. L. Gage, four volumes, Edinburgh, 1866 (condensed from 14 volumes in the original German of the *Erdkunde*).

Part Two

GERMANY

4

Geography after Ritter

DUALISM IN GEOGRAPHY

Geography in Germany during Ritter's later years and for nearly two decades after his death lay under the master's influence. Some of Ritter's pupils produced descriptions of particular areas (*länder-kundliche Darstellungen*), but geography according to Hettner, 'sickened from the one-sideness'[1] bequeathed by the master through his neglect of its physical aspects and overemphasis on man. Thus, 'geography lost its inner balance and its independent significance and sank to the level of an ancillary study to history.'[2]

Ritter's approach was continued by several scholars of no mean distinction. They all had the same objectives, namely, to clearly portray the earth in its relation to man, to establish the influences exercised by the earth as the stage and determinant of human activity on the destiny of individuals and peoples and on the course of cultural development.

The followers of Ritter included (among geographers) Heinrich Kiepert, J. G. Kohl, Ernst Kapp, Hermann Guthe, Karl Neumann, J. Ed. Wappaeus, and G. B. Mendelssohn. Kohl is best known for his early deductive work on the relation of settlements to the earth's surface (1841), and his later work on the geographical location of capital cities (1874). Guthe wrote a textbook on general geography, which in later editions was completely revised by H. Wagner. Wappaeus, who was primarily a statistician, but taught geography at Göttingen from 1838 until 1879, spent the best part of his life writing an encyclopedia of geography and statistics after the manner of Ritter. In 1845 Kapp discussed the distinction between general and special geography (*Vergleichende Allgemeine Erdkunde, 1845*), and wrote broad portrayals of nature and man in major world areas.

---

[1] Alfred Hettner. *Die Geographie, Ihre Geschichte, Ihr Wesen und Ihre Methoden,* Breslau, 1927. p. 88.
[2] *Ibid,* p. 88.

Germany

Mendelssohn described the German lands in terms of their physical build and their peoples, in what Hettner praises as a fine work on historical geography (*Germanisches Europa*, 1836). This was the first attempt to make this kind of treatment. Guthe and Neumann pursued this approach in greater detail, the first in Niedersachsen, the second in Greece, and both their works appeared in 1867.

General geography, as earth science (*Allgemeine Erdkunde*) was pursued outside geography in other disciplines so that the development of a distinctive general geography as conceived by Ritter was impossible. Geology, especially through the work of Charles Lyell, emerged as an organized body of knowledge. Meteorology and oceanography also developed independently. The field of earth science or, as it came to be called, geophysics, began to break way from physics. The study of floral and faunal distributions similarly fell to botanists and zoologists, and their interests only to a minor degree turned to the geographical integration of ecological groups. Similarly, primitive peoples were investigated in the field, and the knowledge organized and interpreted by ethnographers. Geography was thus temporarily neglected and its field empty. The study of the earth's surface forms as an integrated field was divided among the various sciences that sought to examine processes by reference to individual categories of earth-bound phenomena. The concept of natural units (*Länder*) together with their people, as envisaged by Forster and Pallas and developed by Humboldt and Ritter, was neglected by scientists. The best descriptive writings are found in the records of explorers, for this was the heyday of scientific expeditions to distant lands. Some of these travellers, through their experiences, became the new leaders of geographic thought, independent of the Ritter tradition.

Indeed, many geographers in the last decades of the nineteenth century clung to a dualistic concept of geography—on the one hand, the study of the earth as a natural body, and on the other, as the dwelling place of man.[1] Despite the view that a science cannot have two distinct fields but must have a single objective gained increasing momentum, this duality of geography between the two distinct disciplines, physical geography and the geography of man, is a fundamental problem that will always be with us.[2]

---

[1] A review of German works in this period will be found in English in H. J. Mackinder's 'Modern Geography, German and English', *Geographical Journal*, 5 (1895), pp. 367–79. The best appraisals in German are the articles by Hermann Wagner in the *Geographisches Jahrbuch*, Vols. 7, (1878), 8, 9, 10, 12 and 14 (1890).

[2] A critical appraisal of the views of Julius Fröbel (1831) and Georg Gerland (1887) will be found in Hartshorne's *The Nature of Geography*, pp. 72–5 and especially in the section on 'Attempt to construct a "Scientific Geography" ', pp.

Geography after Ritter

A way out of this dualism was sought in the regional concept as the distinctive core of geographic study. This approach, though it had a hundred years of development behind it, did not emerge in a modern sense until the end of the century. It was formulated by Ferdinand von Richthofen in the 1880's and by Alfred Hettner in the 1890's, and was further developed by Vidal de la Blache in the 1900's. Richthofen (see Ch. 6) made his most important statement in 1883, and even this (with considerable justification) was the target of Gerland's basic criticism of the unbridgeable gulf in 1887. Alfred Hettner (Ch. 9) formulated the field of geography in 1905,[1] and explicitly relinquished the field of geophysics (as defined by Gerland) to the earth sciences. This was not a logical definition of the position and status of geography, but essentially a continuation of the tradition of Ritter and Richthofen. However, by declaring geography to be essentially the 'chorological' or 'regional' science of the earth, Richthofen and Hettner sought to resolve the dualism of the field by a 'return to Ritter'. The growth of the systematic fields of geography, and the modes of association of environment and life in particular areal settings have been the substance of geographic research in the twentieth century.

FOLLOWERS OF RITTER[2]

In 1870 geography in Germany was represented at only three Universities—at Berlin by H. Kiepert; at Göttingen by J. E. Wappaeus; and at Breslau by Karl J. H. Neumann. There had also been a lecturer (*dozent*) in geography, later raised to a professorship, at Bonn in 1829, held by G. B. Mendelsohn (a pupil of Ritter), but this became vacant after his death in 1857 and the vacancy was not filled until the appointment of Richthofen in 1877 (until 1883). The big advance took place in the Prussian Universities in the seventies. We begin then, with a short note on the careers of these three scholars who were clearly mature at this date, but, let it be added, their ideas marked the end

---

[1] Alfred Hettner, 'Das Wesen und die Method der Geographie', *Geog. Zeit.*, 2 (1905), pp. 545–64, 615–29 and 671–86. Later included in his book *Die Geographie, Ihre Geschichte, Ihre Wesen und Ihre Method*, Breslau, 1927. (see Chapter 9.)

[2] 'Heinrich Kiepert', *Deutsche Rundschau*, 20, (1889), pp. 569–71; 'Carl Neumann', *Deutsche Rundschau*, 3 (1881), pp. 42–5; 'J. E. Wappaeus', *Pet. Mitt.*, 26 (1880), pp. 110–15.

102–15. Fröbel's article appeared in his twenties and with the approval of Ritter. Through the 1830s Fröbel taught in Zurich and then left academic life altogether. Gerland became Professor of Geography at Strasbourg in 1875. (see p. 94).

of an epoch and had a negligible impact on later developments.

Heinrich Kiepert (1818–99) was the link in the University of Berlin between Ritter and Richthofen. He studied classical history, language and geography in Berlin from 1836 to 1840, under the influence of Ritter, produced an atlas of 'Hellas and the Hellenic colonies' with 24 sheets in 1841 (2nd edition, 1871). He travelled in western Asia Minor at his own cost and produced six maps of that country in 1845 and two on the Turkish Empire in 1844. From 1845 to 1852 Kiepert was at the *Geographisches Institut* in Weimar, but he returned to Berlin to succeed Ritter as dozent of Geography. This post was raised to the status of professor in 1874 (presumably in accordance with the recognition of geography that was taking place in other Prussian State Universities at that time). He was responsible for more maps and atlases of the classical period and wrote a textbook on classical geography in 1879. As was characteristic of his time, in both Germany and France, he taught geography merely as an anciliary to classical history.

Karl J. H. Neumann (1823–80) studied in the university of Königsberg and terminated his junior studies as a student in 1846. He began his career as a private tutor, but was intrigued by political matters and actively participated in the revolutionary movement of 1848. In 1850 he moved to Berlin and served as a newspaper editor for a time, but resumed his academic studies in 1852 with the firm intention of entering a University career. He edited the *Zeitschrift für allgemeine Erdkunde* from 1856 to 1860 and was highly esteemed by both Ritter and Humboldt. In 1860 he became a lecturer at the University of Breslau, but remained in Berlin until 1863 to finish an assignment with the *Staatsministerium*. In 1865 he became a full professor of Geography and Ancient History. His lectures on the history of Rome during the fall of the republic were published posthumously in the early eighties, and a physical geography of Greece, with special reference to the classical period, edited by his successor, J. Partsch, was published in 1885.

Neumann became more interested in geography rather than classical history, although, as in the case of Kiepert, his main training and competence lay in the latter field. He was invited, however, to take over the new chair at Strasbourg (that was finally accepted by Gerland), and he also rejected an invitation to serve as Peschel's successor as professor of geography at Leipzig. He preferred to remain in Breslau and here he died in 1880, where he was followed by his distinguished pupil, Joseph Partsch.

Johann Eduard Wappaeus (1812–79) was born at Hamburg and, for reasons of ill-health, started life by training to be a farmer. In

1831 he went to the University of Göttingen and studied mineralogy. Still suffering from ill-health, he went on a long tropical voyage to the Cape Verde Islands and Brazil. He returned in July 1834 to the University of Berlin to resume his academic studies and was much impressed by the ideas of Ritter. In 1836 he took his doctorate in Göttingen and thereafter continued his studies in Hamburg, Bonn, and Paris. In 1838 he returned to Göttingen as a *privat dozent* in geography and remained in this University for 41 years. In 1845 he became a lecturer and in 1854 attained the rank of full professor.

Wappaeus enjoyed the double title of professor of Geography and Statistics, a frequent combination in the chairs and societies for the advancement of geography at the time. He worked on the geographical studies of Henry the Navigator (1842) and continued his interest in South America that was aroused by his visit of the early 1830s. By the same token, he also published works on German emigration to the New World. In 1847 he undertook the rewriting of a standard 'handbook of geography and statistics', the sixth edition of which had already appeared in 1834. The book was to comprise two volumes and to be finished in two years. Like the master's *Erdkunde* it became a much more extended enterprise. The two books became ten, and the work, with the aid of collaborators, took twenty-two years to complete. Wappaeus also produced statistical compilations, especially of population, gleaned from the new census publications. Throughout his life he clung close to the ideas of Carl Ritter, and stood strongly against the ideas of the critic Peschel (see below), who chose to restrict comparative morphological study of the earth's surface to its physical features.

Although the ideas of Wappaeus had little subsequent impact on geographers, two observations should be made. First he began a long tradition of geography at the University of Göttingen that has continued without interruption. He was succeeded by H. Wagner in 1880, and by W. Meinardus in 1920 who has since been followed by Mortensen, and by the current holder of the chair, H. Poser. Secondly his activity, and that of his successor H. Wagner, reveals the close association of geography with statistics in the early days of the development of both fields of study, and both are associated with the scientific exploration and the emergence of the national census in the second half of the nineteenth century.

OSCAR PESCHEL

Oscar Peschel (1826–75) for some twenty years was the leading academic geographer in Germany. He was basically opposed to the views of Humboldt and Ritter, and devoted the latter part of his life to advancing geographic thought. He was the last great geographer before the full impact of Darwinism was felt, and although he worked after the enunciation of Darwin's theory of evolution (1859), its implications in the interpretation of earth features and human societies were not yet recognized.

He began his career as a journalist, but dedicated himself to geographical writing when still young.[1] From 1849 to 1854 he was assistant editor of the *Allgemeine Zeitung* of Augsburg, and from 1854 until 1870, the editor of the famous weekly *Ausland*, which dealt with foreign countries and affairs. During this period he published books on the age of discovery (1858) and the history of geography (1865), and it was in the latter that he criticized both Humboldt and Ritter. For the last five years of his life he held a new chair of geography at Leipzig University.

Peschel laid the foundation of modern physical geography, through elaboration of the comparative method as expounded by Ritter, with an attempt, however, to explain as well as classify surface features. His work in this connection was published in his *Neue Probleme der vergleichenden Erdkunde als Versuch einer Morphologie der Erdoberfläche* (1869), and in his two volumes of papers, entitled *Physische Erdkunde* (1879), edited after his death by Gustav Leipoldt.

Peschel at various times criticized both Humboldt and Ritter. Humboldt, he asserted gave the impression that 'general geography' was equivalent to the entirety of 'natural science', and Ritter was condemned for his teleological approach and for his subordination of geography to history. Peschel recognized the dualism of geography. He excluded the study of man from it, but devoted his scientific energies and his teaching to both. He made outstanding contributions to the study of the science of physical features of the earth's surface (*physische Erdkunde*) and also made serious investigations of the races and cultures of mankind, which he, like his contemporaries, defined as *Völkerkunde*.[2] Prior to these works, he made a scholarly analysis of the history of geographical ideas down to the middle of the nineteenth century.

In the *Neue Probleme* Peschel criticized Ritter for advocating a

[1] There is a biographical essay by F. Ratzel in *Kleine Schriften*, 1 (1906), pp. 429–47.

[2] Oscar Peschel, *The Races of Man and their Geographical Distribution*, London, 1876, first published in Germany in 1874. It is of interest that this is the only work of Peschel's that is available in English.

method in physical geography which he failed to apply. He argued that comparative geography should have a definite method and aim, like comparative morphology. The geographer should seek, with the aid of large-scale maps, similar physical features in different parts of the earth, compare their characteristics and origin, and endeavour to relate them all genetically, as in comparative anatomy. He deprecated Ritter's teleological treatment which, he claimed, was beyond the sphere of comparative geography, and substituted the comparative study of land-forms, ignoring this influence on human progress, which he considered beyond the scope of the subject.

This comparative method produced valuable results and laid a basis for future study, but his chief difficulty was the lack of knowledge and data concerning the work of the agents of erosion, in particular running water, and the details of regional geology and structure. By studying topographic maps he sought 'homologies', or as Humboldt called them, 'analogies' of form, and then tried to trace their origins. The method was sometimes successful, but many fortuitous comparisons were often made.

An excellent illustration of Peschel's method is found in his essay on fiords in the *Neue Probleme*. From a study of topographic maps he concluded that the essential characteristics of fiords were 'deep and steep gorges in the coasts of continents and islands', which 'frequently extend inland perpendicularly or at a very high angle'. He noted the distribution of such coastal features and pointed out that the feature which 'strongly separates fiords from all similar coastal divisions is their local aggregation and gregarious occurrence', and he concluded that fiords were 'empty dwellings of former ice streams', carved out of fissures due to earth movement. Peschel successfully applied the same method to other physical features, in particular to lakes and islands. He also disproved the widely held idea of the rectilinear arrangement of mountain ranges. In discussing the origin of mountains he pointed out that all the young folded ranges are bounded on one side by land, usually high land, and on the other by an ocean deep, which has sometimes been filled by deposition, an idea taken from Dana's researches.

In an essay on 'Geographical Homologies' there are also examples of the failure of Peschel's method. Peschel noted homologies of form in the mountain framework of Borneo and the shape of Celebes and Halmahera, but he gave no satisfactory reason for their origin. Although he believed the Celebes were probably the skeleton of an old land mass, he could not answer the question whether the three islands are three different forms, or one form at three different stages of development. Again, he noticed the islands of the Pacific are arranged in series like threaded pearls and the similar right-angle bends at the Gulfs of Aden and Oman. The peninsulas and islands to the north of the continent, he noted, are oriented to the north, while on the east and west coasts they trend from north to south, and on the south they taper in the same direction. He discussed the similarities of shape, and found that relief and drainage patterns differed from them. On this basis, he concluded that the continents are older than their mountains, though, in the light of later knowledge, this is now completely disproved, since a continent, is, in fact, a grouping of structural fragments.

Peschel established physical geography in Germany as a science. It had been neglected by Ritter, and Humboldt made no attempt to classify land-forms. Both merely dealt with the lands as a whole, and Ritter's classification was merely based on relief. Peschel, however, attempted to discover distributions and classify them and then to explain the origin of specific land-forms. The method of homologies failed largely because Peschel lacked knowledge of the agents of erosion. Equipped with the researches of the American geologists at the end of the century, and of the *Challenger* and other oceanic expeditions, and with subsequent advances in meteorology, Peschel could have dealt with causes of land-forms and thus avoided fortuitous comparisons. This development, however, was to wait another two decades.

Peschel not only stimulated the scientific study of the forms of the land. He made a definitive contribution to the geographic study of mankind in his book on *Völkerkunde*, that is still widely recognized as an important work by anthropologists. This deals with physical characteristics of mankind (brain, facial structure, stature, skin and hair); linguistic characteristics (history, structure, classification); industrial, social and religious phases of development; and the races of mankind. The last two topics comprise most of the book. The phases of development cover the following subjects—primitive condition, foods, clothing and shelter, weapons, navigation, 'influence of commerce on the local distribution of nations', marriage and paternal authority, social matters such as vendettas, property, slavery, and castle; religious impulses and religious sections in each major religion. The races of mankind are divided (and subdivided) into the following groups: Australians and Papuans, 'Mongoloid Nations' (Tungus, true Mongolians, Turks, Finns, and Samoyeds): 'Dravidian Population of Western India', Hottentots and Bushmen, Negroes, Mediterranean Race (Hamite, Semite, doubtful position of certain European races), Indo-European (Asiatic and European).

Geography after Ritter

Peschel[1] in examining in this book the pre-Columbian cultures of the New World turns to the theme of Nature *versus* Man.

It is very easy, and especially so to us who live in the temperate zone and avoid torrid regions, to recognise the favourable influence on the course of civilisation of the high plateaux within the tropics. Their inhabitants, we say, escape the enervating atmosphere of the sultry lowlands; they were obliged to provide clothing and shelter as a protection against the weather; to avoid starvation, they were obliged to till the ground and to store provisions, and they were even soon forced to congregate and organise societies, in order to meet the requirements of their abode with greater ease. True as all this may sound it does not account for the strange fact that nations should voluntarily have sought out regions in which the difficulties of maintenance were greater. In the Old World, moreover, civilisation was also favoured by the lowlands. It occurs there at the sea level, by the side of the great rivers, such as the Nile, Tigris and the Euphrates. The Chinese too maintain that their civilisation was not developed until they had descended to the Huang-Ho.

THE GENEALOGY OF GEOGRAPHY IN GERMANY

The development of geography in the Universities in the 1870's witnessed the emergence of the first generation of modern geographers. The tradition of Ritter was continued by the three scholars just noted, but their roots lay in classical history and had little to do with the new trends. The situation rapidly improved with the appointment of O. Peschel to Leipzig (1871), A. Kirchhoff to Halle (1873), H. Wagner to Königsberg (1875, Göttingen 1880), F. von Richthofen to Bonn (1877) and G. Gerland to Strasbourg (1875). Wagner and Kirchoff agitated for the inclusion of geography in the Universities and it was largely due to their efforts that the Prussian government decided in 1874 to establish chairs of geography in all the State Universities. It was at this stage that the two leaders entered the academic arena, Richthofen in 1877 at Bonn and Ratzel at Munich in 1875. O. Peschel, who died in 1875 was a third founder of the modern discipline. Other scholars in the last decades of the century were Th. Fischer, J. Partsch, G. Gerland and A. Supan.

In the first quarter of the twentieth century there emerged a considerable number of younger men, all of whom carried out

[1] *Ibid*, p. 444.

prolonged work in the field and this is the period which Van Valkenburg is inclined to describe as 'golden age of German geography.[1]

It is invidious to select from all the geographers of this second generation those who were the leaders. Outstanding, however, were Albrecht Penck and Alfred Hettner, Otto Schlüter and Alfred Philippson. Among these also were Fritz Machachek (1876–1957), who retired from Munich in 1946, and Otto Maull (1887–1957), who retired from Graz in 1945, both of whom were students of Albrecht Penck and were primarily geomorphologists; S. Passarge (1867–1958), a student of Richthofen, who retired in 1935; and W. Volz (1870–1958), a student of Ratzel, Partsch and Richthofen, who retired in 1935. There were two other important contemporaries, both pupils of Richthofen, and both primarily natural scientists. W. Meinardus (1867–1952, a meteorologist), was professor at Münster from 1906 to 1920, and at Göttingen from 1920 (as the successor of Herman Wagner) until his retirement in 1935. Erich von Drygalski (1865–1949), the antarctic explorer, was professor of geography at the University of Munich from 1906 to 1934.

A third generation grew to stature in the thirties and lived their most active years through the Nazi period. Among the most distinguished of the younger men in this inter-war period were Leo Waibel, Robert Gradmann, H. Schmitthenner, N. Krebs, E. Obst, and H. Mortensen. H. Schrepfer and H. Dörries, both of great promise, were killed during World War II.

A fourth generation has grown in the post-war years. Among these outstanding is Carl Troll, but a half dozen other scholars are preeminent—G. Pfeifer, W. Hartke, H. Bobek, J. Budel, H. Lehmann, C. Schott, E. Meynen and H. Schlenger. Both Troll and Meynen state in personal letters their indebtedness to the leadership and ideas of Albrecht Penck. Hartke stresses his relation to A. Rühl in Berlin. The fifth generation embraces the large number of younger men in their late thirties or forties on whom will depend the advancement of geographical research over the next thirty years or so—until 2,000 A.D. No names will be mentioned here, but their work will be discussed, together with those of their French colleagues, in later chapters. This list does not claim in any way to be complete, but it does put into chronological sequence the outstanding makers of modern geography in Germany and Austria.

Germany, unlike France, has not had a single school of thought and training, but several. The most influential in the second generation, to judge from the numbers of their trainees, were Penck in Berlin and Hettner in Heidelberg. Richthofen and Ratzel, leaders of

[1] G. Taylor (Ed.), *Geography in the Twentieth Century*, 1951, pp. 91–115.

the first generation, were contemporaries with an almost identical life span. Albrecht Penck (1858–1945, retired 1926) and Alfred Hettner (1859–1941, retired 1928) dominated the professional scene in the first quarter of this century. We may also note that Otto Schlüter and Alfred Philippson, both students of Richthofen, retired in the thirties but lived to advanced age in the 1950's. They were in the professional sense contemporaries, and these four are but the leading lights among a large and increasing number of professional geographers in the first quarter of the century.

Berlin gave much the same incentive to the modern growth of geography as Paris in France. But other Universities had seats of geography in the 19th century. Bonn, Göttingen and Breslau have a long history of about one hundred years. It is remarkable, however, that Richthofen trained so many young scholars who became the first professors in the chairs that were established in the 1900's, and after his death in Berlin in 1904 he was followed by Albrecht Penck.

5

Leaders of the First Generation: Friedrich Ratzel (1844-1904)

THE STUDY OF MAN

Rapid and far-reaching advances in the study of man were made in the last half of the nineteenth century and these must be appreciated in order to put the development of geography in its proper setting.

A great deal of discussion revolved around the question of the unity of the human races or the independent origins of the various races. In this period, writes an anthropologist reviewing the field, ethnographers were concerned with the classification and description of racial groups. 'The really outstanding classics are O. Peschel's *Völkerkunde* (1873) and Friedrich Müller's *Allgemeine Ethnographic*. The greatest of these was F. Ratzel's *Völkerkunde* of 1885–1888.'[1] Two of these classics were written by geographers who held chairs in German Universities. It may also be noted here that two other outstanding works appeared some years later in Britain—A. H. Keane's *Man: Past and Present* in 1899 and A. C. Haddon's *The Study of Man* in 1898. Both these ethnologists were friends of geography, and their works, as I know from personal experience, were required reading forty years ago for an honours degree in geography in several British Universities.

Arthur de Gobineau wrote in 1854 that the majority of the races were incapable of civilized achievement, no matter how favourable the environment. Following Gobineau, H. S. Chamberlain raised the Teutonic racial stock to a position of leadership in the development of civilization (1899). Georges Vacher de Lapouge, also in 1899, singled out the long-headed Nordic Protestant as the most individualistic and enterprising of the three European races, distinctly superior to the broad-headed Alpine Catholic, obedient

[1] T. K. Penniman, *A Hundred Years of Anthropology*, London, 1935.

62

to government and loath to progress, and the short, dark, long-headed Mediterranean type who was apparently inferior to the other two. Lapouge felt that the superior Nordic type, who was moving to the cities and dying off there, should be preserved by deliberate inbreeding. While this trend led to the establishment of the science of eugenics by Sir Francis Galton and Karl Pearson in London, it also led to popular and pseudoscientific evaluations of the superiority of races, in the early twentieth century (De Tourville, Madison Grant, Lothrop Stoddard, and others), culminating in the racial dogma of the Nazis.

Interest in human populations in their occupance of the earth also led to the study of primitive peoples. The work of Adolf Bastian (1826–1905), a German ethnographer who travelled extensively to carry out field studies in distant lands and who published lengthy ethnographic studies, was outstanding. His main theoretical themes concerned the psychic unity of mankind and the development of distinctive 'folk ideas' in differing environments with migratory movements of ideas between groups. Bastian's theory was set forth in *Zur Lehre von den geographischen Provinzen* in 1888. He suggested that folk ideas spread from particular centres by movements and contacts on their frontiers, and that civilization began through the merger of a variety of such culture streams. He combined both the evolutionary and the diffusionist theories of anthropological study.

The close relations of geography and ethnography at this time are evident in Bastian's being president of the Berlin Geographical Society from 1871 to 1873 and in his active participation in travel and exploration throughout the world. Ritter suggested that the Society should have an ethnographic section in the fifties, and Bastian later established a separate organization for this field while he was President of the Society. He founded and administered the *Museum für Völkerkunde* in Berlin in the 1880's. This close tie of interest is also evident in the two works by Peschel and Ratzel which are recognized as classics by anthropologists. It is also apparent in the interest of other professors of geography in Germany in ethnography and also in the title of the series of research monographs that was begun by Ratzel in 1882 with the title *Forschungen zur deutschen Landeskunde und Völkerkunde*, from which the last part of the title was not omitted until after World War II.

This background of developments in the study of man at the end of the century is essential to understand the progress of geography at that time. Geographers were seeking to secure distinctive objectives such as had been adumbrated by Ritter but not effectively developed. This search applies in particular to the work of Ratzel,

Germany

Richthofen, Hettner, Vidal de la Blache, Jean Brunhes, and Schlüter in the period that reaches approximately from 1890 to 1920.

FRIEDRICH RATZEL

There is no doubt that Friedrich Ratzel has been the greatest single contributor to the development of the geography of man. He also made signal contributions to ethnography that are today highly esteemed by cultural anthropologists. Unfortunately his views have been much distorted by some of his countrymen and grossly mis-interpreted in Britain and America. This has come about mainly through ignorance of the original works in German and through dependence by English scholars on the writings of Ellen Churchill Semple. To find a proper appreciation of Ratzel one must turn to the recent appraisals of two anthropologists, namely Robert H. Lowie and T. K. Penniman, who make it quite clear that Ratzels' work in ethnography and his views on diffusion by migration and borrowing were of outstanding importance. While he systematically sought to select and fix the relations between the physical environ-ment and man, his primary concern as a geographer was with the 'co-variants of human distribution'. Ratzel's main work on ethno-graphy was translated into English with the title *History of Mankind*, and had an introduction by Sir E. B. Tylor, an anthropologist who certainly would have had no truck with naive environmentalism. Moreover, Ratzel's studies of primitive peoples and mapping of their distribution and of their cultural paraphernalia revealed through a historical examination the role of migration and borrowing in their diffusion.

Ratzel used criteria of form. Penniman[1] wrote, 'If he found agree-ment between forms, other than that arising automatically out of the nature, material and purpose of the objects he was examining, he assumed an historical connection or borrowing, even though their forms were widely separated and the distribution discontinuous.'

Out of this, Leo Frobenius, a pupil of Ratzel, developed a theory of culture areas from the recognition of similarities in culture items between Melanesia and West Africa. From the number of similarities between the two cultures he defined the concept of 'criteria of quantity'. This essential procedure has since pervaded the work of ethnologists, notably the American, Clark Wissler, and the approach also prompted the work of the German geographer, Eduard Hahn, on the origin and spread of milking.

[1] *Ibid*, p. 238.

Leaders of the First Generation: Friedrich Ratzel

We need to evaluate this aspect of Ratzel's work rather than to repeat the continued cry of environmental determinism. Most workers during their lives, and especially if they are prolific writers, as Ratzel was, have periodic bursts of enthusiasm during which they oversell their point. Of this Ratzel was guilty. But he urged that consistent man-land relationships exist and that they should be accurately measured. The physiological climatologist, who measures the precise effects of temperature and humidity on the human body and mind, is doing exactly what Ratzel urged should be done. This is neither environmentalism nor naive, even though the results may be inaccurate or distorted. Man-land relationships need to be measured not mystified or muzzled.

The doctrine that a single kind of physical phenomenon always has a corresponding human response through some obscure system of control found its clearest expression in the writing of William Morris Davis. While Davis was basically a geologist (Ratzel was basically a biologist) and though seeking for land-man relationships, Ratzel recognized the limitations of environmental control through human factors. The plain fact of the matter is that Davis's statement of 1903[1] was a geography of land-man relationships and of environmental determinism, whereas Ratzel's was a geography of covariants of human distributions. Proofs of this are found throughout the second volume of *Anthropogeographie*. The central theme of much of his work is that cultural distributions are reflections of migrations and borrowings through history.

Friedrich Ratzel[2] was one of a family of four, whose father managed the staff in the palace of the Grand Duke of Baden in Karlsruhe. After leaving school he served as apprentice to an apothecary for four years. He then took up zoological studies first at the university in Heidelberg, and then at Jena. He published in 1869 a commentary on Darwin's work, that showed strongly the influence of Ernst Heinrich Haeckel, the famous Jena zoologist. While working for a French naturalist he was engaged by the *Kölnische Zeitung* to write popular accounts of his work and travels. After the Franco-Prussian war, in which he fought, receiving several injuries, he briefly resumed study at the University in Munich. Here he met Moritz Wagner, the distinguished naturalist and ethnographer, who was curator of the Ethnographical Museum, and for the first time he came into contact with his theories of the importance

---

[1] W. M. Davis, 'A Scheme of Geography', *Geographical Journal*, 22 (1903), pp. 413–23.

[2] H. Wanklyn, *Friedrich Ratzel; a biographical memoir and a bibliography*. Cambridge, 1961, is the best work in English.

of the migration of species. He then resumed his post with the *Kölnische Zeitung*, which allowed him to travel over much of Europe (especially Austria and Hungary), and many of his writings were subsequently published in a book, *Travels of a Naturalist*.

In 1874 and 1875 he made a long tour of North and Central America. This American experience made a deep life-long impression on him and was the source of many of his ideas. He was perplexed by the Negro problem and impressed by the role of Germans in the development of the American Middle West. His book on Chinese emigration (1876) was prompted by his observations in California, though he used data based largely on British colonial experience. Later in 1878 and 1880, he wrote two volumes on the United States (physical and cultural geography with special consideration of economic conditions). He returned to the theme of the political geography of the United States in a book published in 1893 with special reference to natural and economic conditions. Harriet Wanklyn speaks highly of his vivid descriptive writings in the 1870's and compares them with Humboldt's writings on Mexico and Cuba.

Returning from America, Ratzel resigned from the *Kölnische Zeitung* in 1875 at the age of 31. He became a lecturer in geography at the Technical High School in Munich and was soon promoted to the rank of professor. He remained there until 1886 and became established as an academic geographer. He broke from what he called the integrative morphological approach of Humboldt, who was concerned with the interdependence rather than the origins and spread of things and ideas over the earth. Through his biological approach, Ratzel also departed from Ritter's much criticized approach to geography, but he was far more sympathetic towards it than such contemporaries as Peschel. Ratzel was critical of the mechanistic evolutionary approach of both Darwin and Haeckel. Moreover in his later years, he became increasingly influenced by philosophical rather than biological thinking.

Ratzel gave lecture courses at Munich on a wide range of subjects. He was interested in physical phenomena, such as the level of the snow line and the limestone surfaces in the Alps, as well as in problems of human geography. While at Munich he wrote his *Völkerkunde* (3 Vols. 1885, 1886, and 1888, translated as *The History of Mankind*, 1896–8), two volumes on North America (1878 and 1880), and the first volume of his *Anthropogeographie* or *An Introduction to the Application of Geography to History* (1882). He also wrote frequently for *Ausland* and other papers.

Ratzel followed Richthofen at Leipzig in 1886, when Richthofen was called to Berlin. He remained here until his death in 1904.

Leaders of the First Generation: Friedrich Ratzel

Fischer, Eckert, Friedrich, and Hattner served as his assistants, and he had some eight to ten research students each year. The Leipzig Geographical Society, under his leadership, acquired an international reputation. In these years he wrote the second volume of his *Anthropogeographie*, with the sub-title *The Geographical Distribution of Mankind* (1891), and his enormous *Die Erde und das Leben: Eine vergleichende Erdkunde* (*Earth and Life: A Comparative Geography*) (1901–2), a work that is comparable to Reclus' *Géographie Universelle*. The last work, comments Wanklyn, shows his 'adherence to a physical basis for geographical work', and 'his wide study of contemporary philosophy' and forecasts 'his final concentration on biography' out of which emerged the concept of *Lebensraum*. His final monumental work was on political geography with the title *Politische Geographie oder die Geographie der Staaten, des Verkehrs, und des Krieges* (*Political Geography, or the Geography of States, of Trade, and of War*) (1903). He wrote numerous other articles, many of which are collected in *Kleine Schriften* (2 vols. 1906). A little book on Germany—*Deutschland: Einführung in die Heimatkunde*—is a classic in geographic literature and long served as a standard text in schools. It has passed through many editions since its original publication in 1898, with modifications introduced by Erich von Drygalski in the sixth edition (1932) and by Hans Bobek in the seventh edition (1943).

Ratzel was far from expressing an unadulterated jingoistic pan-Germanism, though many of his more brash statements would lead one to think so. Wanklyn in her recent appraisal writes:

> His abundant writing on contemporary politics shows at times that pugnacity in style and that tendency to speculation which were to lend themselves later very easily to perversion. It also shows an enormous amount of factual knowledge and a sort of rough and independent common sense in judgment which takes him right away from the mystical geopoliticians of the twentieth century.[1]

During his eighteen years at Leipzig, Ratzel had a great influence on the development of geography. He took a leading part, with A. Kirchhoff of Halle, on the 'Central Committee for the Study of the Geography of Germany.' This served as the forerunner of the research monograph series, *Forschungen zur Deutschen Landes—und Volkskunde*, which continues in active publication today and is a prominent feature of geographic research. He also founded and edited a series of works entitled 'Library of Geographical Manuals'. This

[1] *Ibid*, p. 38.

included standard works on climatology by Julius Hann, oceanography by Otto Krümmel, vulcanicity by Karl Sapper, and glaciers by H. Heim. These titles indicate the wide range of Ratzel's geographical framework and the degree to which he encouraged both the intensive study of the homeland and the organization of the varied physical and human aspects of the field.

The above are general comments on the works of Ratzel. Let us now turn to a more specific appraisal of their content and purpose.

The two-volume work entitled *Anthropogeographie* pursues three questions: (1) The distribution and groupings of human population on the earth's surface (this requires maps of different kinds of groups and settlements, especially ethnic, national, linguistic, and religious). (2) The dependence of these distributions on the physical environment and as the result of human migrations. (3) The effects produced by the physical environment on individuals and societies, such as the influence of climate on national character (these aspects were recognized as lying on the frontier of geography).

The first volume (1882) treats the causes of human distributions, that is, the dynamic aspect of geography, and the second, published ten years later, deals with the facts of distribution, that is, the static aspect of geography. The first volume is an application of geography to history, and the second the geographical distribution of man.

Ratzel defined the limits of the ecumene as well as the uninhabitable lands within it, and attempted to account for their frontiers. On the edges of the habitable world were the border peoples, located on the outposts of civilization: Eskimos, Hottentots, Bushmen, Australians, and Tasmanians. He compared the respective locations of these peoples in the northern and southern hemispheres. Migrations within the ecumene were discussed in relation to natural routes and barriers, and the facts which govern man's distribution and development were treated. Climate is used to explain the location of the chief centres of civilization in the temperate belt, and mountains function as frontiers and places of refuge, though rarely were they absolute barriers. Water bodies were one of the greatest obstacles to the movement of primitive man, but these became highways of intercourse when the art of navigation was mastered. The Atlantic, long a barrier, was to Europe and America what the Mediterranean had been to Asia and Europe in antiquity. An analysis of the human geography of coastlines follows. Rivers and marshes prevent expansion, and the latter serve as places of refuge. Forests function similarly, and harbour backward peoples in their clearings.

Ratzel gave special attention to migrations, their classes and causes. He believed every migration had an area of origin and a cause, a route of movement, and a destination. Three sets of 'geographical factors' were said to govern man and his development, namely, situation (*Lage*), involving location with respect to other peoples; space (*Raum*), meaning area which may be either central or peripheral, with a tendency for a people or a state to expand beyond its nursery to its natural limits; and limits (*Rahmen*) being the result of the expansion of neighbouring peoples.

Louis Raveneau[1] in 1892 summed up Ratzel's contribution to geography as follows:

> Between physical geography, sometimes predominant or exclusive, and the science of man, which neglects so easily the framework in which man moves and the space in which he lives, Ratzel has taken his stand. He has strongly insisted on the necessity for a broad view of general conditions and the laws on which depend the distribution of man over the earth. His principal merit is that he reintegrated into geography the human element. By that he has given that science a new orientation and stimulus.

Ratzel's work on political geography, first published in 1897, was an outgrowth of his anthropogeography. The basic concept is that the state is a particular spatial grouping on the earth's surface. By consistent biological analogy, Ratzel regarded the state as an 'earth-bound organism', whereas today we should certainly prefer to use the term 'organization'. Ratzel saw the state as a human group with a definite organization and distribution of life on the earth's surface. Thus, 'every state is a piece of humanity and a portion of the earth'. Out of this arises the concept of *Lebensraum*.

Ratzel believed the state, as a spatial organism, seeks to reach its natural limits. If effective opposition is not offered by strong neighbours, it tends to overflow these limits. 'Geographical, and still more, political expansion', according to Ratzel, 'have all the distinctive characteristics of a body in motion which expands and contracts alternatively in regression and progression. The object of this movement is always the conquest of space with a view to the foundation of states, whether by nomad shepherds or by sedentary agriculturalists.'[2] Human groups and societies always develop within the limits of a natural framework, towards which from a small nucleus they expand and probably overreach, always occupy a definite location on the globe, and are always in need of sustenance.

---

[1] L. Raveneau, 'L'Élement Humain dans la Géographie: L'anthropogéographie de M. Ratzel.' *Annales de Géographie*, 1 (1891–2), pp. 331–47.

[2] F. Ratzel, *Politische Geographie*, Berlin, 1897.

Hence their inevitable association with a definite area, which, with the increase of population, will inevitably expand, until met by natural or human obstacles. These were the three essential geographical facts which govern the character and progress of states.

The book falls into nine parts, which deal with the following topics: the interdependence of land and state, migration and growth of states, the spatial growth of states, the concept of geographical location (*Lage*) in the classification of states, the concept of area (*Raum*), frontiers (*Grenzen*), transitions between land and sea in the spatial development of states, the role of water in the spatial development of states, and the role of mountains and plains in the spatial development of states.

With regard to location, Ratzel referred to the physical location of states as well as to politico-geographical location. Space involved the idea of *total* area as opposed to the *effectively settled* area, and the significance of density of population in the development of the state. He examined the role of the city as a force in creating political areas, and commerce and routes as the arteries of cohesion of the state. Every state sought to make itself into a cohesive geographical entity. Ratzel also developed the concept of the hinterland of a port, recognizing this as falling into five kinds of areal integration—natural hinterland, political hinterland, delivery—market hinterland, products hinterland, and traffic hinterland.[1]

Ratzel first clearly formulated the concept of the cultural landscape. He more frequently characterized it as the historical landscape, since it is a palimpsest of preceding historical phases of human occupance. He urged (as is evident in his classic little book on Deutschland), the classification of fields, farms, villages, towns, and routes with a view to understanding their distribution, their present interconnections and their historical origins. He embraced, as a trained ethnographer, the geographic study of race, language, and religion. He recognized ethnographic groups as geographic assemblies of interrelated phenomena, and sought to characterize the various forms of distribution. He sought to explain them not so much as *in situ* developments of similar phenomena in different parts of the earth, but through the spread and splintering of ideas and phenomena by migrations through history. He formulated the concept of the cultural geographical province as a composite of culture traits reflecting a distinctive group. The notion of cultural spread or diffusion was developed by Ratzel, and he urged use by ethnographers of the geographical techniques of mapping distributions.

[1] F. Ratzel, 'Die geographische Lage der grossen Städte', in *Die Grosstadt, Gehestiftung zu Dresden*, 9 (1902–3), 33 pp.

Leaders of the First Generation: Friedrich Ratzel

This work has had the support and respect of the cultural anthropologists to this day. Ratzel recognized, and rightly so, that land and climate as well as area and location influence man on the earth. This, however, does not make him a naive 'geographic determinist', though many of his statements could lead one to believe so. His approach to geography completely belies this view.

One of the most important concepts of the past generation, *lebensraum*, originated with Ratzel.[1] By this he meant the 'geographical area within which living organisms develop'. He spoke of the general *lebensraum* and the natural *lebensraum* of human groups as a biological habitat. The concept was later used by Karl Haushofer in the wider sense. Heinrich Schmitthenner referred to an 'actual' and a 'potential' *lebensraum*. He referred to it as the 'economic and cultural activity of peoples outside their enclosed settled area'.[2]

I assert here unequivocally that the term *lebensraum*, in spite of its distortion by the Nazis, is one of the most original and fruitful of all concepts in modern geography. It was not designed in the mind of its coiner as a political concept or as a guide to national policy, though, as he emphasized, there was always a tendency for a state to expand or contract its political area according to the measure of its interests or capacities. Ratzel, as a biologist, thought of the anthropogeographic unit as an areal complex whose spatial connections were needed for the functioning and organization of a particular kind of human group, be it the village, town, or state. The concept of *Lebensraum* deals with the relations between human society as a spatial (geographic) organization and its physical setting. Community area, trade area, milk-shed and labour-shed, historical province, commercial entity, the web of trade between neighbouring industrial areas across state boundaries—these are all subsequent variants of the concept of 'the living area'. Ratzel wrote of *Lebensraum* near the end of his life as a fundamental biogeographical concept. Although the idea was abused by the Nazis and their geopolitical protagonists using the works of the Swedish political scientist, R. Kjellen, as a springboard, it was accepted and developed in a scientific sense by the geographers. In its varied interpretation it is a fecund concept of the life and needs of human societies.

Since the views of Anglo-American geographers on Ratzel have

[1] F. Ratzel, 'Der Lebensraum: Eine biogeographische Studie' in *Festgaben für Albert Schaeffle*, Tübingen, 1901, pp. 101–89. See also W. J. Cahnmann, 'The Concept of Raum and the theory of Regionalism', *American Sociological Review*, 9 (1944), pp. 455–562.

[2] See, for example *Lebensraumfragen Europäischer Völker, Band I, Europa*, edited by K. H. Dietzel, O. Schmieder and H. Schmitthenner, Leipzig, 1941. (Also H. Schmitthenner, *Lebensräume im Kampf der Kulturen*, Leipzig, 1938.)

been prejudiced for fifty years and based largely on an ignorance of his work, we quote the balanced appraisal of an informed and distinguished American anthropologist, R. H. Lowie.[1]

> Contrary to some of his expositors, Ratzel did not exaggerate the potency of the physical environment. Indeed, he repeatedly warned against this pitfall and is still further removed from those geographers who see in climate an overshadowing dominant. What saves him from such naivete is the recognition of the time factor. . . . Two further conditions preclude an automatic response to environment: the incalculable effect of the human will and man's unlimited inventiveness. . . . No one could emphasize more than Ratzel the force of past history.

This appraisal may now be summarized as follows. Ratzel's concept of geography involved two approaches: first, the measurement of the consistent interelations of environment and man; and, second the measurement of the interrelations of human phenomena over the earth that are areally coincident. The former focuses on relationships. The latter, contrary to a frequent interpretation, is not concerned with distributions *per se* over the earth, but with the coincidence, correlations, and interrelations of distributions. These are interpreted, therefore, in terms of the physical and the cultural elements that explain such areal interrelations which geographers describe as regional in character. Ratzel also focussed attention on a third aspect, the visible man-made landscape, the *kulturlandschaft*. The appreciation of landscape was the focus of his interest in much of his work and in his aesthetic writings.

The diverse views of geography in Ratzel's work have a common core, but in large measure they pose different problems. It is not surprising that his various conceptual frameworks are reflected in the thinking of his contemporaries. The study of the influence of nature on man remained a constantly reiterated theme in the 'schemes of geography' of such men as William Morris Davis and Hugh R. Mill, who extended their ideas of physiography (like Humboldt) to include the geography of man, though limiting the latter to what was consistently related to the physical environment and derivative from it. This theme appears as the core of the concepts of many others, notably of Vidal de la Blache.

THE HUMANIZED LANDSCAPE

It is natural that the assessment of the influence of environment on

[1] R. H. Lowie, *The History of Ethnographical Theory*, New York, 1937, p. 120.

man should have its counterpart in man's influence on nature as an agent of conquest and change. Indeed, the character and process of the human imprint on the landscape becomes even more sharply focussed as one of the main objectives of geographic inquiry. Without making reference to earlier authorities, we again refer to the writings of Count Buffon in *Epoques de la nature*, where he remarked on the landscape made by man. 'Men destroy woods, drain marshes and lakes, and in process of time, give an appearance to the surface of the earth totally different from that of uninhabited or newly peopled countries.' Buffon also cited instances of climatic changes brought about by forest clearing and noted how man had been able to re-fashion the earth by the use of domesticated plants and animals. Although Buffon sometimes upbraided man for his destruction of nature, he was, on the whole, optimistic. His notion of the epochs of nature was of a sevenfold sequence in the history of the earth in the last of which 'the power of man assisted the works of nature'.

Further interest in conservation was aroused by studies of the floods of Alpine torrents and the changes in level of Alpine lakes that were attributed by N. T. de Saussure to the cuttings of the woods in the headstreams of the rivers. Humboldt also made similar observations. These and other observations in the early nineteenth century led to the belief that the destruction of forest and the spread of cultivation caused climatic change. Civilization leads to aridity, was the theory. Such ideas antedated the later work of Petr Kropotkin, Raphael Pumpelly, and Ellsworth Huntington. Outstanding was the work of Victor Hehn on *The Wanderings of Plants and Animals*, (1885), who wrote, 'the whole physiognomy of life, labour, and landscape in a country may, in the course of centuries, be changed under the hand of man.' Hehn instanced the transformation brought about by man in the Mediterranean lands by the diffusion of plants and animals brought in from southwest Asia, but urged that 'the depradations of the goat, deforestation, and soil erosion need not be irreparable'. The geologist, Charles Lyell, continued the theme. He considered man as 'among the powers of organic nature' that modi-fied the surface of the earth, and in terms of its inequalities, served as a levelling agent.

The notion was carried a big step further by George Perkins Marsh. This American, basing his observations largely on European experience and literature (he served on official missions in Turkey, Greece and Italy) wrote an extended work on the subject in 1864 with the title *Man and Nature: or, Physical Geography as Modified by Human Action*. Its purpose, he wrote, was 'to indicate the charac-ter and, approximately, the extent of the changes produced by human action . . ., the dangers of imprudence and the necessity of caution in all operations which, on a large scale, interfere with the spontaneous arrangements of the organic or the inorganic world'.

In this work, writes Glacken, the technical ideas of the nineteenth century scholars 'converge with the seventeenth century conception of nature as a divinely designed balance and harmony', and they sought 'the restoration of disturbed harmonies'. Marsh examined man's sway over the earth in terms of its impact on animal and vegetable life, and on woodland, waters, and sand, but the effects of deforestation received a major share of the treatment. His notions of the betterment of human adjustments included international migrations, the balance of nature, revolutionary spread of cultivated plants and domesticated animals, such as goats and camels as agents of modification, effects of war and fashion on the land, climatic changes due to deforestation, drainage, reclamation, irrigation, fixation of dunes, and the conservation of ground water.

Marsh was a man of high reputation and some of his ideas in matters of land settlement were adopted. But his warnings were lost in the optimistic mood of the Victorian epoch, that was voiced by men like Herbert Spencer who talked of the 'march of civilization as new lands were populated wherein equilibrium would be established and the world cultivated like a garden'. Marsh's plea for conservation was not really effectively heard until the twentieth century. But, writes Glacken, 'Our greatest debt to him is that he studied the technical works of European foresters, meteorologists, agronomists, drainage engineers and hydrologists, botanists and plant geographers and scientific travellers and for the first time placed the results of their investigations where they belonged—in the forefront of human history.'

Other investigators observed such phenomenon and emphasized them in their writings. Élisée Reclus, for one, developed the theme, factually as well as philosophically. The German geographer, Ernst Friedrich, also turned to the various forms of exploitation of the land involved in the spread of the European peoples. In *Petermann's Mitteilungen* in 1904, he coined the term *Raubwirtschaft* or robber economy. Such an economy passes through three phases—intensive and prolonged exploitation, followed by impoverishment which leads to an awareness of the need for conservation. This sequence of experience, Friedrich wrote, led to the use of soil fertilizers in Europe, and 'we see before our eyes processes going on in the new lands whose completed results may be observed in Old Europe'. The destruction of animals, plants, and soil is the way of progress. This robber economy was a phase of colonization and would be brought to rational practices under the leadership of Europeans. These interesting ideas were widely adopted in the early 1900's and have played an important part in the assessment of the relationship between man and land.

Reference should be made here to an article by Alexander Woeikof in 1901. He was appalled at the growth of cities. 'This grouping in cities, under conditions unhealthy to body and mind, this dissociation of man and the earth is proof of a sickly state.' Pointing to the way in which the destruction of forest and grassland permitted soils to be disintegrated by wind and rain, he urged the careful control of surface vegetation so as to restore and maintain the harmony of nature and to hold the soil in place. Otherwise, he pointed out, we are hurtling headlong to the destruction of civilization. If conservationist steps were taken, he envisaged 'vast perspectives of progress', including great potentialities of the inter-tropical lands that, he thought, could hold some ten billion people.

Recent experts on soil erosion have been equally pessimistic in their assertions in a world that is still not actively aware of the need and gains of soil conservation. Man's role in changing the face of the earth received little attention in the nineteenth century. Environmental determinism became predominant through the operation of the Darwinian evolutionary doctrine, though it is in direct contradiction to what was going on in an age of technological progress and ruthless exploitation of the pioneer lands of the earth. The idea of man as an agent of geographic change and of the need for preserving or reestablishing the balance of nature and maintaining or increasing its productivity, which is the essence of conservation, did not begin to take root among scholars and practitioners until well into the twentieth century.

Man as an agent in the transformation of the surface of the earth has become in the twentieth century a central theme of investigation among geographers. The conservation of natural resources has become a matter of increasing concern, especially in the new lands of the world, though there is still a wide gap between the theorist and the practitioner.

Further References

In addition to the works of Ratzel that are listed in the text, attention is drawn to two recent studies: J. Steinmetzler, *Die Anthropogeographie Friedrich Ratzel's und ihre Ideengeschichtlichen Würzeln,*

Germany

Bonner Geog. Abhandlungen, Heft 19, 1956, and Harriet Wanklyn, *Friedrich Ratzel: a biographical memoir and bibliography*, Cambridge, 1961. Hermann Overbeck, has two articles, dated 1952 and 1957 reproduced in his Kulturlandschaftsforschung und Landeskunde, *Heidelberger Geog. Arbeiten*, Heft 14 (1965) pp. 60–103.

Singularly little attention is given to the works of Ratzel in Hartshorne's *The Nature of Geography*.

6

Leaders of the First Generation:
Ferdinand von Richthofen (1833-1905)

Freiherr Ferdinand von Richthofen was born on May 5th, 1833 at Karlsruhe in Silesia, a member of a noble family of that province. Strongly attracted to geology, he began his researches in the Alps and later under the auspices of the Austrian government, in the Carpathians. He studied the granites and dolomites in the south Tyrol, and interpreted the latter as coral formations. This was a new idea which Albrecht Penck used later in his studies of Alpine geology.

In 1860 he accompanied a Prussian expedition to eastern Asia, and then went to California where he stayed for six years. He was always interested in the nature of the relations between volcanic rocks and the occurrence of gold, which he had studied in Hungary, but he had also practical interests. He became a journalist and informed the German public through his newspaper reports of the wealth in gold of California. He warned against the investment of capital in gold-mining undertakings, but recommended the establishment of a smelting works with German capital. He reported from San Francisco on the famous Comstock lode. He seized upon his great ambition in 1868—research in China, the geology of which was quite unknown. For this undertaking he was first financed by the bank of California, and later by the Chamber of Commerce of Shanghai who financed four years of travel throughout China so that he might report to it in English (of which language he had perfect command) on the economic conditions of the country, in particular the coal resources, the great importance of which he was the first to draw attention. In the scientific field, he studied geological structure and land forms. As for his professional obligations, he supplied the Chamber of Commerce with detailed reports that form a veritable quarry of information on the economic geography of China which are exemplary studies in that field.[1]

[1] *Letters from Richthofen, Shanghai*, 1870–2.

Germany

He had the idea of writing a major work on China, in which he would present his scientific investigations. His homeland offered this opportunity. In 1872, he returned to Germany after an absence of twelve years. The new Reich had just come into being, and the government was lavish in its grants in aid of research. Richthofen received government grants for his work on China, that was to occupy intermittently the rest of his life. The fact that the work was incomplete at his death, was in large measure due to the onerous duties that were piled on his shoulders in Germany in his new academic life.[1] I here draw attention to the content of this classic work.

In the first volume Richthofen deals with the structure of the mountains of central Asia and their influence on the movements of peoples. He discusses in detail the gradual growth of knowledge of China. He recognizes the "loess" deposits of northern China as wind blown dust from the steppe. This whole approach, he described as a geographical experiment; and his work gave definite direction to the development of modern geography.

The second volume is concerned with north China, and is based mainly on his observations. Geology and land-forms receive chief attention, for, as he maintains, research into the inner and outer structure is the basis of knowledge of any and every land. But he also considers the people and their activities, and includes much of the economic material that appeared in his letters to the Shanghai Chamber of Commerce.

The third volume was to deal with south China on similar lines, but he died before achieving it. The work was completed from his materials by Ernst Tiessen. A large atlas of China was also to accompany the work, but Richthofen did not finish this, and Max Croll worked on the maps of southern China after his death. These works were published at the expense of the Prussian *Kulturminister-ium*.

The city of Berlin made many demands on a man so well travelled, so well informed and with such acute judgement. He was quickly placed at the head of the *Gesellschaft für Erdkunde* which he raised to the first rank among geographical societies. The Prussian Government lured him to the service of the University of Berlin in 1875, although he was at once granted leave of absence to continue his writings. He was called to the University of Bonn as Professor of geography in 1877, so his whole energies were now placed in his teaching and the China work lay in abeyance. He was soon after

[1] Ferdinand von Richthofen, *China: Ergebnisse eigener Reisen und darauf gegründte Studien*, Berlin, 1877–1912, five volumes and atlas.

78

called to Leipzig in 1883, and there devoted himself to the development and presentation of a clearly defined system of geography that profoundly influenced geographic thought in Germany.

Leipzig has a long tradition in geography. In 1871 Oscar Peschel was appointed as the first Professor of Geography. This great scholar was already near the end of a long editorial career and died in 1875. The University was then confronted with the question of his successor. There followed an 'interregnum' of nearly ten years, and it was not until 1883 that Richthofen was appointed. In 1885 he declined an invitation to Vienna which Albrecht Penck accepted. He moved to a new chair in physical geography in Berlin in 1886. On leaving Leipzig, Richthofen suggested Penck as his successor. Penck however, declined since at that time he preferred to stay in Vienna. Consequently Friedrich Ratzel a specialist and contributor in an entirely different field, accepted as Richthofen's successor and remained in Leipzig until 1904. It is of interest to note that twenty years later Penck decided to accept the invitation to Berlin where he succeeded Richthofen.

In 1886 Richthofen was called back again by the Prussian Government to the University of Berlin to fill a newly established chair in geography, that aimed at the pursuit of geography proper, alongside historical geography that was already pursued by a second professor (H. Kiepert) in the University.

Richthofen was not universally accepted in academic circles in Berlin. He was still regarded by many primarily as a geologist, but the geologists themselves, who were inclined to think only of their own land, found his approach too wide-sweeping. It was long before he found his due place in the *Akademie der Wissenschaften* in Berlin, but his personality and abilities gradually prevailed. He raised questions in his lectures that had never been touched in the circles of the University. His pronouncements on the geography of settlement and transport were published posthumously (edited by Otto Schlüter). He drew many mature students to his *Kolloquium*, some of whom became travellers, like the Swedish Sven Hedin, and others became teachers in schools.

When Richthofen first came to Berlin in 1872 the great phase of German colonial expansion in Africa was at hand. When he returned to the capital in 1886 Germany held colonies in Africa and in the South Pacific, so that another field was opened for exploration. Richthofen was true to China. His dear wish was that Germany should obtain a footing in the Far East. He had long before in his book on China drawn attention to the strategic importance of the bay of Kaiuschau as a gateway into northern China. That the Reich eventually found a footing here was due in large measure to him. He always advocated closer bonds between Germany and China, for he greatly admired the Chinese civilization. He was, however, unable to raise funds for further exploratory work in China, for Africa remained the chosen field of the German people at the time. At the end of the nineteenth century Germany turned its attention increasingly to the sea and to the expansion of its navy and so there gradually developed the idea of a marine museum in Berlin. Richthofen was persuaded to undertake this task. He did not desire a propaganda institute for the navy, but a scientific institute and a museum for the enlightenment of the general public. He gave the last years of his life to this project; and as Rector of the University, he delivered a special lecture on this theme, but he did not live to see the completion of the task, for he died shortly before the *Museum für Meereskunde* was opened.

Richthofen was of an enquiring disposition with a sharp sense of problem. His *Führer für Forschnungsreisende* (1886)[1] urged greater depth in research and drew attention to many open questions, among which, of course, as already shown in his work on China, the relations of geological structure and surface features loomed very large. The book, indeed, as a guide to scientific explorers, was a morphology of the earth's surface features.

The first part deals with the techniques of field observation, especially in its relevance to physical geography and geology. It embraces the preliminary choice of route, maps, literature, etc., and then the measurement, mapping, and the collection of meteorological records.

The second is the main part of the book and is in fact the first systematic interpretation of the processes involved in the shaping of the earth's surface. It covers: observations on the process of mechanical weathering; the ground water and springs; the mechanical work of running water; continental water bodies; sea-coasts and islands observed on sea-voyages. The work of running water includes the work of a river, in terms of ablation, transport, corrosion and deposition; and the variations of lithology and structure, tectonics and climate that affect this normal development of the river bed and its valley forms. In each case the land-forms are classified in terms of the processes.

The third part deals with observations of the soils, rocks and mountain structure. Here chapters cover observations of the soil (soil forming factors, distribution of soil types); observations of rocks (crystalline, sedimentary, eruptive); observations on volcanoes;

[1] *Führer für Forschnungsreisende*, Berlin, 1886.

the structure of mountains; and the principal kinds of land-forms (*Bodenplastik*); and observations of useful minerals.

The land-forms are classified as to type (and often regional distribution) under the headings of the various processes. Thus, for example, the development of a river is portrayed, with profiles in terms of the mass of water and the vertical gradient and the aberrations from this normal development due to other factors— structure, lithology, earth movement, vegetation, exposure, and climatic region.

The differential development of a river system and the associated kinds of relief are examined in areas of unconsolidated strata, horizontal strata, and abrasional surfaces. Also in relation to processes involved (earth movements and surface erosion and deposition), lakes, coasts and islands are classified in the appropriate chapters.[1]

If we ... regard Mountain (*Gebirge*) and Lowland (*Flachboden*) as the two basic kinds of surface form, so several modes of approach may be used, in order to establish further divisions within these two classes, or ... certain types. This has been attempted especially with regard to mountains. For a long period a distinction has been made between mountain ridges (*Kettengebirge*), that have a decided long axis, and mountain masses (*Massengebirge*) that do not have such a form. It appears better to use the genetic factor and to subdivide mountains into types according to the basic forces in the development of their outer forms. But since various forces operate in the development of every mountain system, as in other divisions in the preceding pages, we have taken in every single case as a basis of division that force that has been primary in shaping the form. ... Then secondary divisions can be made on the basis of secondary forces. But one cannot, as in most other fields of physical geography, arrive by this method at sharp distributions; since the forms merge into each other and at times two different forces operate equally together. As a last resort one can use the external forms as the basis of sub-classification.[2]

Social connections permitted Richthofen to exercise a wide influence in both scientific and diplomatic circles. These connections he exploited to encourage scientific exploration, to which end he

---

[1] It is important to note that the interpretation of river action is based on German, French and American works that *preceded* the published results of W. M. Davis' investigations.

[2] *Führer für Forschungsreisende* pp. 640–641.

won the support of the *Gesellschaft für Erdkunde*. Thus he took an active part in promoting and organising a south Polar expedition under the leadership of Drygalski, whom he esteemed highly and who later became one of the leading German geographers. It was while reporting on this project that he died suddenly in 1905 at the age of seventy-two.

Richthofen has rightly been compared in his significance for geography with Ritter and Humboldt. He shared with the latter his long period of exploratory work in the field. But while Humboldt's researches led to the contemplation of the total *Kosmos* and he ended his days in prosperous and leisured seclusion, Richthofen limited himself to the study of the earth's surface and became an active teacher in his later years. His geography was very different from that of his predecessor, Carl Ritter. While Ritter depended primarily on the writings and observations of others, Richthofen's work, like that of Humboldt, was built upon his own observations through years of travel. While Ritter drew packed audiences to his public lectures, Richthofen exercised his influence on small groups in the more intimate atmosphere of his *Kolloquium*. Thus, he trained many young scholars who carried on his work, whereas Ritter left no one of outstanding quality to continue his work. Richthofen regarded geography as the study of the earth's surface (*Erdoberfläche*), and bridged the gap between geology and geography. But this is not his only strength. His letters to the Chamber of Commerce at Shanghai contain work on economic geography of outstanding quality, and his interpretation of the Treaty of Shimoneski published in Hettner's new periodical in 1895 is a masterpiece of political geography.[1]

We may now turn to Richthofen's views of the problems and methods of geography. Richthofen gave his inaugural address at the University of Leipzig on *Aufgaben und Methoden der Heutigen Geographie* in 1883 and the following paragraphs are a summary of its content. Geography, he said, is the science of the earth's surface (*Erdoberfläche*) and the things and phenomena that are causally interrelated with it.[2] It is not 'Earth Science' (*Erdkunde*), for this would be comprehensive. He preferred *Erdoberflächenkunde*. Its methods are measurement and observation of phenomena in the field. Geography may be pursued through the most detailed investiga-

---

[1] *Siedlungs—und Verkehrsgeographie*, Berlin, 1908. Posthumous publication, edited by Otto Schlüter.

[2] 'Die Geographie gestaltet sich dadurch zu der Wissenschaft von der Erdoberfläche und den mit ihr in ursächlichen Zussammenhang stehenden Dingen und Erscheinungen'.

tion of the smallest areas, as well as through the comparative study of larger areas. Thus there are two approaches according to whether the areas or the things and the phenomena are the primary object of study. The first is Special Geography and is primarily descriptive. The second is General Geography. One is synthetic, the other is analytical. The combination of both methods yields a third approach that considers selected groups of things and phenomena in a particular area and seeks to understand their interrelations and causes. This is the chorological approach.

Descriptive, synthetic or special geography (*Beschreibende Geographie* or *Spezial Geographie* or *Beschreibende Erdoberflächenkunde*) he defines as follows:

> Every area on the earth, no matter how large or small, whether a continent, a small island, or a naturally bounded inland area, an artificially bounded state, a mountain, a river basin or a sea, is examined as a grouping of smaller unit areas, as well as in the perceptible appearances, among which are included the works of human culture.

It has two methods of approach. First, the earth surface (*Erdoberfläche*) consists of component areas and the whole can be pierced together by the juxtaposition of these areas. Second, any area of the earth's surface (*Erdraum*) is an agglomeration of things and phenomena that are elements of the six realms of nature and can only be portrayed through consideration of them all. 'The scope of chorography, as so defined, is comprehensive, eclectic, and inexhaustable.' It is encyclopedic, a *Reportium des Wissens*.

Thus, there are two levels of descriptive geography which he describes as chorography and chorology. Chorography does not go beyond the systematic assembly of all the appearances of the individual land areas (*Erdräume*). The number of these is so great that one is obliged in practice to follow an 'eclectical procedure'. The description on the human side of a small area on chorographic lines will include the distribution of population, races, language, frontiers, settlements, industries, religions, trade centres, routes and products. The number of items is unlimited into which the chorographic study may reach, and when man is considered these must be extended to include the historical aspects. Thus, synthesis is the keynote, although an analytical method is first required in breaking down the whole area into its component parts. The chorological approach is made possible by the advancement of various contributing sciences. It is concerned not only with registering the area facts that are there, it also attempts to explain the areal distribution of these phenomena through the introduction of causative and dynamic interrelationships of the phenomena on every single portion of the area's surface (*Erdraum*). Strabo was the forerunner of this approach. This was the mode of treatment of Ritter and Humboldt. Richthofen notes, however, that Ritter handled only in a fragmentary way the areal interrelationships of natural phenomena upon which Humboldt laid so much emphasis, since he (Ritter) regarded the assessment of the influence of the physical forms—natural resources, water supply and plant cover—upon man as the highest goal of the chorological approach. This chorological approach has been facilitated since Ritter's time by the growth of the special disciplines that deal with the explanation of the spatial distribution of particular sets of phenomena. This leads to the field of General Geography (*Allgemeine Geographie*). The more deeply a study goes (into process rather than distribution), the less use is it to a chorological treatment of a particular area.

Richthofen turns next to the abstract or analytical method. This is termed *Allgemeine Geographie*. The mode of presentation of chorography is didactic and proceeds from the areal distribution of facts and conditions to their interrelations and causes. Synthesis dominates, though the analytical process is explored in the breakdown of the whole area into its parts. General geography, on the other hand, is not *progressive*. It is rather *regressive*, since it passes from the particular to the general, from the condition (*Bedingten*) to the cause (*Ursächlichen*).

General geography studies earth-bound phenomena on a fourfold base—the forms; the material (*Stoff*); forces and causes of change (*Kräfte*); and movement (*Bewegung*). These four points of view lead to the morphological, material, dynamic and genetic modes of approach. The fourth moulds the science, since it examines retrospectively the effects of the earlier forces upon the earlier phenomena, that is, their evolution from a previous stage. One can thus divide General Geography into distinctive aspects according to these four principles, or one can use the last (genetic) approach as the basis of interpretation of the other three. Richthofen prefers to study each of the six realms of nature under the heading of these four principles.

The main goal of geography is to establish the interrelations of phenomena in areas. Humboldt pursued this method. Ritter put a setback to it through his philosophical bent, so that General Geography (*Allgemeine Geographie*) fell into separate camps such as geology, botany, zoology. Richthofen states:

Only what was related to ethnography, such as customs, laws and clothing of peoples, remained to geography. Since Man was the goal and purpose of Ritter's geographical work, research into the nature of areas could only be handled by special sciences, and geography was merely used as a means to an end.

History and geography were thus regarded as sister sciences. But soon the latter sank to the service of a hand-maid to the first and only modest place was found in it for ethnography.

Under these conditions both analytical geography and chorology became impossible. Much of the integrating research in physical geography in the early nineteenth century (physical, biological, and ethnographic) was lost. This, says Richthofen, was probably because of the lack of clear definition of geography as an academic discipline and for this Ritter is mainly responsible. Ritter's chorographic presentations were stimulating in his day. They were based on a grandiose philosophical conception, that failed to present a tangible research method. Richthofen sought to revive the concept of unity of the earth's surface and to bring the analytical approach into a closer relationship with chorological study. Developments in the individual sciences in the mid-nineteenth century and their methods of approach now permitted the ideal of Ritter to be applied and seek to present the facts of terrestrial unity. The various branches were being brought once again in Richthofen's day within the sphere of geography as is evident in the work of such men as Oscar Peschel and Élisée Reclus. Geography (says Richthofen) now links the problems of the various sciences in a unity, the basis of which is the earth's surface (*Erdoberfläche*) and the things and phenomena areally arranged on it.

After discussing the content and method of general physical geography, Richthofen turns to consider the geography of Man. This is concerned with the distribution of races and tribes, peoples and nations, languages and religions, the causes of which are extremely complicated and are by no means dependent only on biological laws. In this field there are also four modes of approach, namely, the outer form or aspect (*äussere Gestaltung*), interrelations (*Zusammensetzung*), causal relationships (*ursächliche Beziehungen*) or dynamic forces (*Kräftewirkungen*), and genetic development. The basis of the first is morphological. The second includes race, religion and language and the ethnical principles of grouping. 'The relations of all these categories of phenomena to the earth's surface as a whole and in its parts are subject to constant change and development.' Hence the need for what Richthofen calls a 'dynamic anthropogeography' which aims at 'the understanding of the influence of the nature of the lands (*Natur de Erdrüme*) on Man, as well as the influence of Man on the transformation of the nature of lands'. This is the field to which Ratzel has contributed (says Richthofen) and from which Richthofen hopes to see the emergence of a system of laws of causal relationships.

The genetic approach, he continues, is close to history. It is concerned with the relation of man to the natural environment. Agriculture, irrigation, mineral exploitation, industry, settlements constitute the static elements of geography, whereas trade and commerce make up its dynamic element. All these are concerned with man's material culture. On the other hand, there are the evidences of his spiritual culture (*geistige Kultur*) in their local occurrences and development, as well as in their spread from one people to another and from their seat of origin to wider areas.

In summary—still according to Richthofen—geography is concerned with the six realms of nature—the fixed forms of the earth's surface—the lands, the hydrosphere, and the atmosphere: the plant and animal world; man and his material and spiritual culture. These are each studied from the four points of view: form (*Gestalt*), material content, spatial interrelationships, and development. The guiding principle throughout is that of the interrelation of the three realms to each other and to the earth's surface.

This concludes the summary of Richthofen's famous address.

Richthofen was probably the greatest and most effective contributor to the development of modern geography. He trained many young men, some of whom became distinguished geographers and geologists. Among these were the Swedish explorer, Sven Hedin, and the two exchanged letters in a long correspondence. Siegfried Passarge studied under the master. So also did Otto Schlüter, who was responsible for collecting and editing his lectures on the geography of settlement and trade. Alfred Philippson studied under Richthofen at Leipzig, and was one of his oldest students, whereas Alfred Rühl (see Ch. 11) was his last doctoral candidate in Berlin.

The geographers who followed the Richthofen direction fall into two groups in the first quarter of this century, all of whom had their base in a firm grounding on the physical aspects of the earth's surface and especially in geology. The first and older group handled the distribution of categories of phenomena either on a country base or on a world-wide base. The second and later group took particular areas and presented an orderly arrangement of the categories with their base in the physical (geological) environment.

The first group dominated until World War I and was directed mainly to advances in the natural sciences. It was concerned with seeking general principles of distribution. The various categories—land, air, water, plants and man—were considered separately in their relation to the physical earth. Outstanding contributors were A. Penck and F. Ratzel. To this group there also belonged other contemporaries—A. Kirchhoff, A. Supan, Th. Fischer, and A. Philippson. The approach is most clearly demonstrated in Penck's book on the *Deutsche Reich*, published in 1887. Two other works in this period were Supan's book on Austria-Hungary, published in 1889, and Theobald Fischer's book on southern Europe, published in 1893.

The second, and younger group, sought to establish the inter-relations of spatially arranged phenomena within the framework of physical units, this being the method used by Richthofen in his own work on China. Its outstanding contributors were Joseph Partsch and Alfred Hettner. The major work by Norbert Krebs on *Länder-kunde der Osterreichischen Alpen*, published in 1913, follows the same procedure and is a standard of its kind. To these outstanding scholars there should be added the names of F. Machatschek, R. Gradmann, Erich von Drygalski, M. Friedrichsen, O. Maull, and G. Braun. There is evident in the works of this group the pursuit of particular problems of geographical distribution in particular areas, and all of them, in the Richthofen tradition, were strongly based in geology and geomorphology.

Richthofen's system of geography was an assembly of spatial distributions, that could be treated either with respect to the character and principles of general world-wide distribution, or with respect to the various sets of phenomena as they occurred in one particular area. In the latter case, the primary foundation and determinant was considered to be the 'geological-morphological base', and the other spatially distributed phenomena were presented as a combination sandwich or a storeyed structure or, as H. R. Mill conceived it, as a pyramid, with a physical base and an economic rubble heap at the top. A contribution to the knowledge of any particular area demands special skills in field and library, in order to collect, collate, map and interpret the relevant data. Some, like Gradmann, were able to overcome this problem by specializing on some particular fields, but these difficulties of expertise were enlarged by Richthofen, who urged his students (like Philippson) to work in areas that were little known and unmapped. And, let it be emphasized, there are still vast areas of the globe where precisely such reconnaissance studies, presented as organized descriptive inventories, are still urgently needed. But, however thoroughly such a series of layers may be presented and mastered, when put together they offer little new or unique, for there lacks a thematic enquiry, that, with its own questions, selects the relevant data from the various levels of the sandwich or pyramid. The repercussion of this criticism will be evident in what follows in the twentieth century.

Further References

This chapter is mainly based on the following sources:

A. Hettner, 'Ferdinand von Richthofen's Bedeutung für die Geographie', *Geographische Zeitschift*, Volume 12, (1906), pp. 1–11.

E. von Drygalski, 'Von Richthofen und die deutsche Geographie', *Zeit. Ges. f. Erdkunde Berlin*, (1933), pp. 88–97.

G. Wegener and H. von Wissmann, 'Ferdinand von Richthofen', in *Die Grossen Deutschen*, Berlin, 1935–6, Volume 5, pp. 390–8.

F. von Richthofen, *Aufgaben und Methoden der heutigen Geographie, Akademische Antrittsrede*, Leipzig, 1883.

7

Contemporaries of the First Generation

JOSEPH PARTSCH

Joseph Partsch (1851–1925) is ranked in a recent German appraisal with Hettner and Penck in the growth of modern Geography in Germany.[1] His professional career coincides almost exactly with theirs, for it ranged from the early seventies to the early twenties. He was definitely a leader of the first generation and his activity and influence continued through the first quarter of the twentieth century.

Partsch, unlike both Penck and Hettner, who both began their careers, like Richthofen, in the natural sciences, grew out of a firm foundation in philology and classical history. Through his close association at Breslau with Carl Neumann, as well as with H. Kiepert in the chair of geography in Berlin, he had a direct link with the approach of Carl Ritter, who was the teacher of both these men. Yet Partsch steadily broadened his approach to geography by his interest in his homeland, in the field, in the library, in the map, and eventually in print. He was an eager walker, a keen observer of nature and landscape, and an enthusiastic sportsman and mountain climber. He emerged as a great contributor to the advancement of regional geography, notably in the areas of Silesia and central Europe. His published books and monographs, spread over a period of fifty years, cover 36 works and articles and other items number

---

[1] H. Overbeck, 'Joseph Partsch's Beitrag zur landeskundlichen Forschung', *Berichte für Deutschen Landeskunde*, (Jan., 1953) pp. 34–56. Also an appraisal by A. Penck in *Zeit. d. Ges. F. Erdkunde zu Berlin*, (1928), pp. 81–98. The fullest treatment is H. Waldbauer (Ed.), *Joseph Partsch; Aus Funfzig Jahre: Verlorene Schriften*, Breslau, 1927. This contains a biography (pp. 7–34) and a speech on the development of geography in the nineteenth century in Germany, on the occasion of his becoming Rector of the University of Breslau in 1899. The work on Silesia is entitled *Schlesien: Eine Landeskunde für das deutsche Volk auf wissenschaftliche Grundlage*, 2 volumes, I, *Das Ganze Land*, 1896; and II, *Landschaften und Siedlungen*, Heft 1, *Oberschlesien*, Heft 2, *Mittelschlesien*, 1907; and Heft 3, *Niederschlesien*, 1911.

more than 160, together with several hundred reviews of books.

Born in 1851 in Silesia in the Riesengebirge, he was the son of a family of seven, his father being the manager of a glass factory with a considerable reputation. He moved to Breslau in 1860 to live with friends of the family and to attend the gymnasium, and in 1869 he entered the University of Breslau to study classical philology. Among his teachers there, he was most impressed by Carl Neumann. Neumann, a student and devotee of Carl Ritter, after some years in Berlin, accepted a professorship of 'Ancient History and Geography' in 1860, and remained there until his death in 1880. He gave his first course on geography in 1863, and thus continued a study that fell into decline in the years after Ritter's death. Partsch attended Neumann's course on 'general hydrography' in the autumn of 1869, and thereafter turned to geography under the sympathetic guidance of his master. In 1871 he won a prize for an essay, established by Neumann, on 'Chorography, Topography, and Administrative History of Ancient Africa'. It reflected the approach to geography as cultivated by Neumann and Kiepert. An extension of this work earned the doctorate in 1874. In 1875 Partsch compiled a study of the concept of Europe in the works of Agrippa, for which he passed the *habilitation* for 'Geography and Ancient History'. Shortly thereafter he passed the State examination, which qualified him to teach five subjects—Latin, Greek, History, Geography, and Education. In November, 1876, he was appointed in the University as *ausserordentlicher Professor*. He was now able to support himself, whereas in the previous years he had received help from the family or earned money by private teaching. His attention was devoted in these early years to classical history and philology. But he soon became interested in the natural sciences, especially geology and geophysics, which he submitted at the *habilitation* examination. His inaugural address in October 1875 dealt with the significance of Silesia for Prussia and Germany and this revealed the beginnings of a geographic approach. His philological interest, however, persisted through his life. This is revealed by a review of Kiepert's *Lehrbuch der Alten Geographie* and his publication on the works of a Byzantine poet, sponsored by Mommsen. He also made a study in 1891 of the contribution of Philip Cluver in the seventeenth century to historical regional geography (*historische Länderkunde*), published in a series edited by Penck.

Partsch was an eager devotee to observation in the field and was a walker from his early days in the Riesengebirge. Here he developed an interest in glaciation and glacial land-forms, and extended his observations to the High Tatra and to the Alps. He visited the Alps several times in the late seventies and was particularly observant of glaciers in the Riesengebirge in 1878, and observed moraines and gravels that spoke of extended glacial activity in a past period of time. These observations he extended to the High Tatra and in 1882 published a book with the title *Die Gletscher der Vorzeit in den Carathen und in den Mittelgebirgen Deutschlands*, a work that appeared almost simultaneously with A. Penck's book on *Die Vergletscherung der Deutschen Alpen*.

In 1880 Partsch decided to concentrate on geography and abandon classical history. Neumann died in that year, and Partsch set about editing and supplementing his unfinished work on the physical geography of Greece, published in 1885, that followed the ideas of Ritter. This work was undertaken before Partsch had ever been to Greece. In the following years, however, he visited both the mainland and the Aegean islands, under the auspices of the Berlin Academy and supported by H. Kiepert of Berlin. Out of these visits emerged several monographs, notably on Corfu and Leukas, that appeared in the separate volumes of *Petermann's Mitteilungen*.

Partsch was always deeply interested in the land and people of Silesia and in the procedures of regional geography. In 1887 he became a member of the central commission on the 'scientific regional geography of Germany' (established by Ratzel a few years earlier), with special responsibility for Silesia and Posen, to which the kingdom of Saxony was added in 1905. In 1889 he committed himself to the publication of a bibliography of literature on Silesia, and this was printed in seven parts between 1892 and 1900 with the title *Landes-und Volkskunde der Provinz Schlesiens*. In 1889 he published a school text, *Eine Kleine Landeskunde der Provinz Schlesien* (small geography of Silesia), a modest forerunner of his later work, that went through eight editions, the last in 1918. Monographic studies of the glaciation of the Riesengebirge (1894) and the rainfall of Silesia and bordering lands (1895) were published in the research series of the *Forschungen zur deutschen Landes—und Volkskunde*. The first, as we have noted, and must emphasize, had occupied his attention for over ten years. The second was based on a map, prepared from the records of 527 stations.

In the late nineties H. J. Mackinder approached Partsch to write a regional geography of central Europe, as one of a series of twelve envisaged by Mackinder in a series on *Regions of the World*. Partsch, after some hesitation, accepted the area defined by Mackinder as central Europe, since this seemed to be the area that had long been dominated by German culture, and it also was given distinctive character by the threefold association of Alps, Central Uplands, and

Northern Lowland. Partsch completed the work during the period 1897 and 1899, though it was not published until 1903, almost simultaneously with the first volume of his work on Silesia. The English translation was abridged so as to conform with the series. Partsch, after reworking it, had the whole of the original manuscript published in Germany in 1904 with the title *Mitteleuropa: Die Länder und Völker von den Westalpen und dem Balkan bis au den Kanal und das Kurische Haff.*

The work on central Europe and the first volume of the work on Silesia gave to Partsch an international reputation and renown in Germany. He was invited to Leipzig to succeed Ratzel, and although he had previously refused offers to go to Königsberg (1885), Vienna (1902) and Halle (1904), he accepted this one and thus painfully severed a life-long connection of 44 years with the city of Breslau. He accepted what Neumann had refused three decades earlier, when the latter was invited to succeed Oscar Peschel in 1875.

On arriving at Leipzig in 1905, Partsch immediately completed his work on Silesia, and the second part on *Mittelschlesien* appeared in 1907, and the third part on *Niederschlesien*, in 1911. Partsch's continued interest in the classical period is still evident in his inaugural address on Egypt's importance for geography, and in a study of Aristotle's book on the rising of the Nile published in 1909.

In the Leipzig period he made two extensive foreign journeys. He attended the International Geological Congress in Stockholm and travelled in northern Europe (1910). He was invited by the American Geographical Society to the United States in 1912 and participated in the trans-continental excursion.

He was invited to Kiepert's chair in Berlin in 1914 but declined, yet he served for many years as president of the *Gesellschaft für Erdkunde* in Berlin, until 1921. He was greatly respected by the government of Saxony and served as its representative at the discussions on the international map in London in 1909 and Paris in 1913.

The first world war caused Partsch to turn his attention to relevant matters. He addressed an audience of 5,000 in the main hall in Leipzig on Germany's eastern frontier, and he interpreted the course of the war in various theatres. He also turned to matters of political geography in the years immediately after the war.

In 1921 the twentieth meeting of German geographers took place in Leipzig, the first since the war, and the first since the Strasbourg meetings of 1914. (Partsch had been the principal figure in the meetings held in Breslau in 1901.) In 1922 he retired after nearly fifty years of professional service. But shortly thereafter he published a work on the High Tatra, with the title *Hohe Tatra zur Eiszeit*, that included a map of glacial forms on a scale of 1:75,000, profiles, sketches, and many photographs and was concerned with the land-forms produced by glaciation of past periods. In his closing years Partsch was occupied with a work on the geography of world trade. This was completed by R. Reinhard and published posthumously in 1927. Joseph Partsch died in June, 1925.

Partsch did not build up a large school of doctorate followers, either at Breslau or Leipzig. He brought his assistants from other Universities and preferred not to employ his own trainees. Among his assistants, almost all of whom attained considerable academic distinction in other Universities, were the following—E. Friedrich, A. Merz, J. Sölch, W. Behrmann, O. Lehmann, E. Scheu, H. Neubert, and H. Rudolphi. The influence of Joseph Partsch's geographic scholarship was also felt indirectly among his colleagues. Norbert Krebs wrote as follows: 'I count myself in this respect with pride as a pupil of Partsch, although I never had the good fortune to sit at his feet'. In almost the same words Partsch recognized his debt to H. Kiepert, the classical geographer at Berlin and a close friend: 'Many who never sat at his feet recognise him as their master and were aware that from him they learned and were furthered on their own way.' This speaks conclusively of the geographic tradition from the founder Ritter, through Kiepert and Neumann, through Partsch, and then to Krebs. The development of the discipline from its over-emphasis of classical geography to a newly conceived and applied field of regional geography is evident in the career of this one man, a great scholar of his time.

HERMANN WAGNER

Hermann Wagner (1840–1920)[1] was the son of Rudolph Wagner, a well known physiologist, who was professor at Erlangen and came to Göttingen in 1840 as the successor of Johann Friedrich Blumenbach, the famous anthropologist. The young son was brought up in Göttingen as his home town. He studied mathematics and physics in these same Universities, and this field was the first objective of his academic career. He accepted a post at a gymnasium in Gotha to teach geography, mathematics and natural history and held this

---

[1] W. Meinardus, 'Hermann Wagner', *Nachrichten der Ges. d. Wissenschaften zu Göttingen, Gesellschaftliche Mitteilungen*, 1919–30. pp. 60–5. W. Meinardus, 'Hermann Wagner', *Pet. Mitt.*, 75 (1929), pp. 225–9. Bibliography, 1864–1920, in *Pett. Mitt.*, 66 (1920), pp. 118–22.

from 1864 until 1876. He became increasingly interested in promoting the teaching of geography in schools. In 1868 he became the editor of the statistical section of the famous *Almanach de Gotha* in the Perthes establishment. In 1866 there was published (in collaboration with E. Behm), *Die Bevölkerung der Erde*, as a volume of *Petermanns Mitteilungen*. This was essentially a statistical compilation for which he had particular competence. In 1876 Wagner was called to a new chair at the State University of Königsberg where he started an institute from scratch on his own. He brought out a revised edition of Guthe's *Lehrbuch der Geographie* (4th ed., 1877–9 and 5th ed. 1882–3). In its fifth edition (1883) this book was entirely rewritten by Wagner and contained new parts on general geography (*Allgemeine Erdkunde*) of extra-European countries, and a regional geography of Europe (*Länderkunde von Europa*). This went through various editions and modifications thereafter reaching a final tenth edition in 1920. Wagner undertook the editorship of the *Geographisches Jahrbuch* in 1879 and held this post for forty years until his death. The 36 volumes that passed through his hands recorded the development of all aspects of geography, and in their pages he wrote long and important articles on the scope and method of geography in research and education. He was also a prime mover in the formation of the meetings of professional geographers (*Geographentag*), the first of which was held in 1881, since which date the meeting has been held every two years. In 1880 with the death of E. Wappaeus in 1879, Wagner accepted the chair at Göttingen and remained there until 1920.

During his forty years at Göttingen many young geographers studied under him for varying periods. Among those who entered academic careers were W. Sievers, W. Behrmann, F. Mager, and H. Dörries. Colleagues (*dozenten*) who worked under him were O. Krummel, M. Friedrichsen, L. Mecking and F. Klute.

GEORG GERLAND

Georg Gerland (1833–1919)[1] was professor of geography at Strasbourg from 1875 until his retirement in 1910. His views have had little repercussion on the development of geographic scholarship in Germany, but the strange fact is that Alfred Hettner took his doctorate under Gerland, not because he agreed with the views of

[1] See R. Hartshorne, *The Nature of Geography*, pp. 89–90 and 106–20, and Alfred Hettner's autobiographical notes in *Heid. Geog. Arbeiten*, Heft 6. (1960), pp. 41–81.

Gerland, but because he found the society at Strasbourg congenial and stimulating. Gerland was originally a classical philologist and was a pupil of Theodor Waitz, the distinguished anthropologist. Gerland taught in school for many years and was appointed to the new chair of geography at Strasbourg in 1875 a few years after the foundation of the University. He was then 42 years old and had no previous training in geography. He tried at first to convert the chair to one in ethnography (*Völkerkunde*), but this effort failed. He then sought to organize himself as a geographer and turned to the views of Kant and O. Peschel in an effort to develop geography as an exact science. He threw his activities entirely on the side of the physical aspects, and advocated the complete exclusion of man from the field of geography. He established a new periodical, the first of its kind, in geophysics and it was in the pages of the first number in 1887 that he supported the view that geography as the earth science should be exclusively concerned with the physical earth. Gerland seems to have had real difficulty in adjusting his attitude to the heritage of geography. His radical views seem to have had little impact on his own research or teaching in his many years at Strasbourg. But, says Hartshorne, his views are significant, if abortive, in that they sought to develop a field of scientific geography on the basis of logical principles, regardless of its historical development. A long discussion of Gerland's position, and particularly of the thorough rebuttal by H. Wagner in the *Geographisches Jahrbuch* for 1888, will be found in the *Nature of Geography* by Hartshorne,[1] who concludes:

> Granted the need for a study of the physics of the earth, we need only recognise how completely different such a field is from that which scientists as well as laymen have for centuries called geography. Gerland's thesis, as completed in the discussions then and since, may well stand as the *reductio ad absurdam* of all attempts to make geography into either an 'exact' or an 'essentially natural' science.

Hettner tells us,[2] from his personal aquaintance with Gerland, that the latter regarded geography and ethnography as two distinct fields. Hettner says that Gerland was a great *polyhistor*, but never became a geographer. His lectures, says Hettner, were collections of information and his excursions were 'eclectic'. Hettner had cordial relations with his senior, though he could not agree with his attitude to geography and especially his views on the geography of man.

[1] Hartshorne, *Ibid*, p. 115.
[2] *op. cit.* footnote 1, p. 94.

Germany

Hettner completed his work on the climate of Chile under Gerland. It is ironical that Hettner, who became a great geographer, should have taken his professional training under Gerland, who raised a storm of protest from his colleagues and whose views have had no influence on his successors.

THEOBALD FISCHER

Theobald Fischer (1846–1910)[1] was the *erste Privat dozent der Geographie*', says H. Wagner. For in 1876 he took his *habilitation* in geography at the University of Bonn, then moved to Kiel in 1879, and finally reached Marburg in 1883 where he remained until his death in 1910. Fischer started his studies in history and then transferred to geography. He suffered from ill-health as a young man and spent a long period of convalescence in the Mediterranean, whence he derived his life-long dedication to the geography of that area.

Fischer became a recognized authority on the Mediterranean lands. His first work was on the physical geography of this area particularly of Sicily (1877) and culminated in a monograph on field work in the Atlas foreland of Morocco (*Pet. Mitt.*, 1900, *Erg. Heft.* 133), An Italian translation of his book on the geography of the Italian peninsula, described as a 'work in scientific chorography', appeared in 1902. He was also responsible for thorough studies of the date palm (1881) and the olive (1904), that can be classed, according to Wagner, as exemplary works on the model of Ritter. Fischer is best known for his monumental work on the geography of the Mediterranean lands (*Länderkunde der südeuropäischen Halbinseln*), that was published in 1893 in the Kirchoff series on *Länderkunde von Europa*. He intended to write a second volume on the north African lands associated with the Mediterranean. He was plagued throughout his professional life, at home in Marburg and on his field excursions in the Mediterranean countries, by ailments of the heart and lungs and died at an age when he still had plans for further research and writing.

ALFRED KIRCHHOFF

Alfred Kirchhoff (1838–1907)[2] was the oldest and first *ordinarius*

[1] Obituary notes by H. Wagner in *Pet.Mitt.*, 56, (1910), pp. 188–9 and A. Rühl in *Geographische Zeitschrift*, 27 (1921), pp. 29–38.
[2] H. Steffen, 'Alfred Kirchhoff', *Geographische Zeitschrift*, 25 (1919), pp. 289–302.

professor for geography in the Prussian Universities. He held his post for more than thirty years and turned out several thousand students who became teachers in schools. He was specially interested in the training of teachers, since this is how he started his own career. He laid special emphasis on *Länderkunde* and dealt only superficially with general aspects. He was a voracious reader, a masterful lecturer and teacher, and a well loved personality among the student body. His writings were also dedicated mainly to the needs of the schools, as for example his series *Erdkunde für Schulen*. He also made a habit of giving many lectures outside the University. He served for a long time as the editor of the *Forschungen zur deutschen Landes-und Volkskunde*, and edited 14 of its volumes. He was for two years (1871–3) a lecturer at the Military Academy of Berlin, but was called to head a new institute of geography at the University of Halle in 1873. Here he remained until retirement in 1904, so that his period of professional activity synchronizes almost exactly with that of Richthofen and Ratzel.

Kirchhoff influenced several men who attended his lectures, and became distinguished geographers. We have already named Hettner and Schlüter, who began their studies under him. Schlüter was diverted from Germanistic studies to geography. Hettner was attracted to methodology by Kirchhoff, though he criticized his teacher for being a 'popularizer' rather than a 'scholar'. Other pupils of Kirchhoff, who presumably completed their training under him, were R. Credner and A. Supan. It is of more than passing interest to recall that Carl Ritter studied at the University of Halle at the end of the eighteenth century, so that Halle has very old and respected associations with geography.

Kirchhoff was also the editor of a large regional geography of Europe (*Länderkunde von Europa*), published in 1886 with collaborators in Th. Fischer, F. Hahn, A. Penck and A. Supan. He followed closely the conceptual framework of Richthofen, based on the natural sciences.

ALEXANDER SUPAN

Alexander Supan (1847–1920)[1] is regarded by H. Wagner as one of the great *Autodidakten*—self-made geographers—of the 1870's. He was, in other words, one of the first generation of geographers in Germany, although he regarded himself as a pupil of A. Kirchhoff.

[1] H. Wagner, 'Alexander Supan', *Pet. Mitt.*, 66 (1920), pp. 140–6 and B. Dietrich in *Geographische Zeitschrift*, 17 (1911), pp. 193–8.

He was born in the Pusterthal and went to school in Laibach and studied literature and history at the University of Graz. He took his doctorate in history in 1870 and then moved to a teaching post in an elementary school in Laibach. In the early seventies he began to take an interest in geography and wrote a book on geography for schools in 1873. In 1876 he wrote an article on the content of geography for the transactions of the Geographical Society of Vienna, and in this he voiced his opposition to the school of Ritter and sought to establish geography as a 'pure natural science'. He obtained leave of absence from his school and went to study under Kirchhoff at Halle and also to continue his geological studies. He returned to Laibach in 1877, but his leave was renewed, largely through the support of Kirchhoff. He now went to study at Leipzig, but made frequent visits to Halle. He moved to the *gymnasium* in Czernowitz in 1877 and shortly thereafter in 1881, took his *habilitation* in geography.

In Germany he became known through his *Lehrbuch*, which was largely rewritten in its third edition in 1878. In the eighties he turned to questions of the measurement of land-forms and meteorological phenomena. His work on the thermal zones of the world was published in 1879 and 1880. He agreed with a publisher to write a book on physical geography (*physische Erdkunde*) and this was first published in 1884.

In the year 1884 occurred the death of E. Behm, the co-founder of *Petermanns Mitteilungen* and he was followed, largely on the final advice of H. Wagner, by Supan. Thereafter, for nearly twenty-five years Supan directed and built up the reputation of this outstanding periodical, a period that corresponded with the academic careers of Richthofen and Ratzel. Supan stated his objectives clearly when taking over this post. 'The days of exploration', he said, 'are essentially over, and we shall dedicate the *Mitteilungen* and the supplementary monographs to the collection of data and the advancement of *Landeskunde* as a scientific field of knowledge.' He also undertook the editorship of the *Statistisches Jahrbuch* of Gotha. He vastly expanded the reviews of books, for most of which he was personally responsible and he contributed articles of substance on climate in the 1890's. A large number of the supplementary monographs passed through his hands and the majority were on *Länderkunde*.

His own scholarly work continued apart from the editorship of the *Mitteilungen*. He contributed a large section on Austria-Hungary to Kirchhoff's *Länderkunde von Europa* and this work, says Wagner, puts him in the front row of German geographers. He also collected statistical data. He devoted a monograph to the role of the states of the United States to world trade (1886) and renewed the publication of population statistics that had been begun by E. Behm and H. Wagner. The development of 'colonial areas' was also of concern to him and he wrote a book on the territorial development of the European colonies in 1906. He thoroughly rewrote his physical geography (*Grundzüge der physischen Erdkunde*) in 1896, a work that passed through a number of editions until the last in 1916, and was long regarded as a standard work.

He was offered the chair at Breslau, in 1910 and accepted it at the age of 62. He suffered, however, from ill-health for a long period, but even so he wrote a new work on political geography (*Leitlinien der allgemeinen politischen Geographie*) that was published in 1918. In this book he pleaded for the development of a political geography independent of the dictates of the natural sciences. This he regarded as his most original work. He died in 1920.

8

Albrecht Penck (1858-1945)

The geographers of the second generation were active until around 1930, certainly before the onset of the Nazi era. Most of them were pupils of Richthofen, or at least studied under him, and these included Penck at Leipzig, Hettner at Bonn, and Schlüter at Berlin. Others were Erich von Drygalski (1865–1949) who retired in 1934; K. Hassert (1868–1947), a student of Ratzel and Richthofen, who was professor (after short spells at Leipzig, Tübingen and Köln) at Dresden from 1917 until his retirement in 1935; Siegfried Passarge (1867–1958) at Hamburg who retired in 1936; Alfred Philippson (1864–1953) at Bonn who retired in 1929; Wilhelm Volz (1870–1958), who studied under Ratzel, Partsch and Richthofen, and became Professor at Erlangen in 1912, Breslau in 1918, and Leipzig in 1922 until his retirement in 1935; and Wilhelm Meinardus (1867–1952), a physical geographer and primarily a climatologist and meteorologist, who succeeded to the chair at Göttingen in 1920, (where he had served since 1909) in succession to Wagner, and who remained there until his retirement in 1935. To this second generation there also belong Karl Sapper and Robert Gradmann, who were members of no particular school but entered, as it were, from outside on their own scholastic merits, the first as a widely travelled expert on Central America, with a firm base in the natural sciences, especially geology, and the second as a specialist on South Germany with a strong training in botany.

The dominant leaders of this second generation were Albrecht Penck, Alfred Hettner and Otto Schlüter, and to their life and works we shall first turn our attention.

Albrecht Penck was without doubt one of the scholastic giants of the last generation. His greatest contributions to knowledge were his studies of the ice age and its fluctuations, but he was throughout his life a professional geographer. In addition to his monumental work on the Ice Age in the Alps he wrote, in the eighties, an important work on the geography of Germany and a large classic work on the morphology of the earth's surface in 1894. His published works extend over a period of sixty years.

Born in Leipzig in 1858, Penck studied natural sciences at the University of Leipzig, beginning in 1875. He went to Munich in 1880 to work under the geologist Karl Zittel and became a *privat dozent* at the University of Munich in 1883, at the age of 25. After two years he was called to the chair of physical geography at the University of Vienna. He remained there for nearly twenty years, and built up a well equipped geographical institute. During this period he carried out his greatest work on the fluctuations of the Ice Age, based on observations of the deposits in the valleys of the Alps. In 1906 he accepted the chair of geography at the University of Berlin, as the immediate successor to Ferdinand von Richthofen. This move he made at the age of 48 and he retained the headship for twenty years (and also served for a year as Rector of the University in 1917–18). He retired in 1926 and was immediately succeeded by one of his outstanding Austrian pupils, Norbert Krebs. On retirement Penck continued to live in Berlin where I had the privilege of meeting him several times in 1936. During the war he was bombed out of his home and moved to Prague, and there he died in March 1945, at the age of eighty-seven.

Albrecht Penck had not reached his twentieth birthday when he was chosen by the geologist Herman Credner for work on the geological survey of Saxony and produced the geological map of an area southeast of Leipzig on the scale of 1:25,000. At this time he developed not only his skill of observation, but also the interpretation of maps as a record of the features of the earth's surface. 'Nicht die Karte, sondern die Natur war ihm die erste Quelle'.

He became interested in the problems of the Ice Age early in his career, and stayed in this area as his primary research field, for sixty years, as he states in an autobiographical note. His birthplace, Leipzig, lay near the southern edge of the ice sheet. In 1879 he published his first paper, which was on the boulder clay (*Geschie-beformation*) of the lowland of Germany. Various hypotheses were current at this time to explain the character and distribution of surface deposits and topographical forms in the European lowland; and the 'drift theory' of transfer of rocks on the backs of icebergs from Scandinavia was still widely held. This was gradually displaced by the 'ice sheet theory' to which Penck now accumulated masses of evidence in support. Penck found near Leipzig under the *Geschie-belehm* (boulder clay) a polished rock (*Rundhöcker*) of Scandinavian affinity. He sought widely for further evidence in northern Europe, and finally brought out the paper noted above in 1879. He developed the theory further in a book (*Vergletscherung der deutschen Alpen*) in 1882. This work was Penck's entry (*habilitation*) to geography at the University of Munich.

The theory of climatic fluctuations in the Ice Age was not new. Penck's contribution lay in the remarkable collection of his field observations to support and develop it. The alternation of deposits of boulder clay (*Geschiebelehm*) and gravels led him to postulate at least a threefold invasion of north Germany by a Scandinavian ice-sheet. In his 1879 paper he writes 'We are dealing with glacial and interglacial periods. Each glacial period of the Ice Age is represented by a *Geschiebelehm*, and each intermediate interglacial period by laminated (*geschichtete*) sands and clay masses.' This is the basic idea derived from evidences in the northern lowland of Germany. The next problem was to find out whether this was supported by evidence in other parts of Europe. He sought this in the Scottish highlands, Alps and Pyrenees. His major field work was undertaken in the Alps, based on Vienna, and his great work, undertaken jointly over a period of twenty years with Eduard Brückner—*Die Alpen im Eiszeitalter*—appeared in three volumes from 1901-9. This was the culmination of his personal field researches. His division of the quaternary Ice Age in the Alps into three interglacial and four glacial periods (named after Alpine rivers, Günz, Mindel, Riss and Würm) became the classic basis of quaternary geology and the springboard for the interpretation of human pre-history. This is the great contribution of Penck to glaciology and quaternary geology.

The comparative method of field study practised by Humboldt was basic to Penck's work in the field—observe, locate, compare—in respect to a particular problem. The phenomena observed—deposits, cuttings, sequences etc.,—must be precisely localized, mapped and put together in their spatial occurrences, with respect to physical quality, depth, thickness and extent.

His objective was thoroughly geographical; his method to that end was geological. From the operation of exogenetic processes and morphological forms one can often deduce evidences for the development of crustal movements. (This approach, like that of W. M. Davis, was attacked by S. Passarge.) The objects and methods of this study are portrayed in his work on the morphology of the earth's surface, published in two volumes in 1894.

The *Morphologie der Erdoberfläche*, a fundamental work and the first of its kind, embraced the 'morphology of the earth's surface'. In discussing its scope, Penck distinguishes the study from Geodesy and Geophysics and uses the term Geomorphology. In its concern with the forms it is geographic. While German scholars in the nineteenth century always referred to this as the study of the 'morphology of the earth's surface', it was Penck, so far as we can discover, who first used the term geomorphology.

Penck's work fell into two volumes. The first volume dealt with the metric measurement of earth forms and the processes involved in the formation of the earth's surface. The second covers the individual geographic complexes of similar and related forms, classified and examined in terms of their mode of development. This involved a careful empirical description of the various types of land surface. Such a study, Penck points out, was first undertaken by O. Peschel in his study in comparative geography that bore the sub-title of 'the morphology of the earth's surface' (1869), while the studies of the agents of erosion and their impact on the interpretation of landforms, he continues, were put together by the French geologists, de la Noe and E. de Margerie, in 1888. Noe's work was being rapidly undertaken at the time of Penck's writing. Davis made his impact in the 1890's and 1900's, his lectures in Berlin and Paris making a widespread impact so that in the last edition of Penck's work there were substantial changes in content. The basic structure, however, remained unchanged.

In the first volume he considers at length the agents of erosion and deposition—weathering, mass movement, river action, aeolian action, glacial action, and endogenous forces. In volume two the land-forms are considered at length under several main descriptive headings—plains; hill country (dunes, morains, volcanic and sinter hills); valleys and their relation to various forms of composite relief—dissected and open, remnant hill groups, tableland, residual hills; basins and lakes (in deserts, glaciated areas, areas of porous rock (*Karst*), areas of deposition, volcanic mountains (cuesta, faulted, folded and volcanic); depressions (*Senken*); hollows. The forms of the sea are considered under the headings of coast lines, sea floors and islands.

It will be noted that Penck gave especial emphasis to the description of the forms of earth features—morphometry (or morphography) as well as to the processes of formation (*Kräftelehre*), and to the locale and grouping of similar *forms* into distinctive regions. His system of classification is based on *form* not on *process*. Thus, for instance, *Wannenlandschaften* included a great variety of forms that were derived from a variety of processes—glacial, karstic, volcanic, depositional, etc. It would be absurd to claim that Penck underestimated as a geologist, the role of process in the development of land-forms. He logically insisted that in the geographical study of land-forms the primary emphasis should lie in the descriptive

103

measurement and classification of the forms of the land and their grouping, from the smallest topographic units and contiguous groupings into distinctive areas of different orders. He distinguished six *topographic* forms or *form elements*:

1. The plain or gently inclined uniform surface.
2. The scarp, or steeply inclined slope.
3. The valley, composed of two lateral slopes inclined to a narrow strip of plain which itself slopes down in the direction of its length.
4. The mountain, a surface falling away in every direction—which may be a point or a line (ridge).
5. The hollow, the converse of four.
6. The cavern or space entirely surrounded by a land surface.

These topographic forms do not occur alone, but are grouped together into *Landschaften* of different orders—districts, regions, and lands. The character of the topographic surface depends largely on the structural surface. The six chief *structural forms* are plains, with horizontal strata; slightly folded strata (*Verbiegungsland*); faulted blocks (*Schollenland*); intensely folded areas (*Faltungsland*); lava overflows (*Ergussland*); and intrusive volcanic masses (*Intrusivland*). With the action of the agents of erosion, and particularly running water, on the structural forms are produced the topographic or land-forms. On this basis Penck worked out his classification.

As a geomorphologist, Penck sought the origin of land-forms in genetic processes through time, as an anatomist examines an organism. Recent morphological analysis has shifted its emphasis to functional processes, as a physiologist examines the organism. Thus, climate is considered by younger workers to have a more significant part to play in explaining the existing character of land-forms than the relation between the geological structure and the surface forms as was emphasized by Penck and Davis. Both were concerned in understanding the present surface forms.

A formidable list of distinguished scholars worked under Penck during his earlier years in Vienna. These included R. Sieger, N. Krebs, F. Machatshek, H. Hassinger, A. Merz, J. Sölch, all of them Austrians. J. Cvijic and E. de Martonne were among his foreign students. Penck was outstanding, with Richthofen in Berlin, in the German speaking lands.

It is important to name the geographers who trained under Penck, or were strongly influenced by him, during his twenty years at Berlin—the later half of his career. These include Gustav Braun, Norbert Krebs, Emil Meynen, Carl Troll, Herbert Louis, Herbert

Lehmann, and his earliest assistant and today the dean of German geographers, Hermann Lautensach.

Penck took over the direction of the Institute and of the *Museum für Meereskunde*, that had been sponsored by Richthofen. The museum received his special attention and he was involved in the oceanographic expedition of the *Meteor*. He also encouraged the Antarctic expedition of E. Drygalski. He himself worked on climatic classification (1910 and 1913) and he used the terms *arid, humid* and *nival* with specific physiographic connotations, as the basis of a world classification of climates. In the twenties and thirties he continued to work on geomorphological problems in the field in the Inn Valley and on the correlation of human occupance with climatic changes in the Ice Age. Later he turned to the question of the population capacity of the earth, in terms of the carrying capacity of specific land units. This was the subject of several years of intensive work by many students in the Institute under Penck's direction. Its results are summarized in a book, that came out in 1942, and must be one of his last writings.

He turned to questions of political geography during and after World War I and especially to the range of German culture in Central Europe, drawing a distinction between *Sprachboden*, *Volksboden* and *Kulturboden*. These were new ideas and stimulated much critical work that the exaggerations of the Second World War obscured.

> *Deutschland* is for us a natural entity (*eine natürliche Einheit*) and not simply a political conception; such a one is the *Deutsches Reich*. This latter is the state with its frontiers, which unfortunately do not embrace all those who would belong to it. *Deutschland*, however, is also not the land in which German only is spoken. It stretches from early times beyond the limits of German speech (*Sprachboden*), and does not coincide with the latter. What *Deutschland* has meant for centuries is a definite part of the earth's surface, with characteristic features, and a distinct form (*Gestalt*).

This was written in 1925 and crystallizes a problem that is reiterated in numerous studies by German scholars. It embodies refinements of Ratzel's important basic concept of *lebensraum*.

This view of Penck was re-echoed by his successor, N. Krebs in 1929 who wrote as follows:

> *Deutschland* is for us neither a pure physical, nor a pure cultural (*völkischer*) conception; and it certainly does not coincide with the changing area on the political map. It is, as the name indicates,

Germany

the land that is German on the basis of all the physical and cultural peculiarities of the area. *Deutschland* is greater and more stable than the *Deutsches Reich:* but also its area is not unalterable for all time.

Krebs was expressing views shared with and probably acquired from Penck. The viewpoint is evident in the magnificent atlas of the German *Lebensraum in Mitteleuropa* that came out in the thirties. In this atlas the *German Lebensraum* is treated as the area of German settlement and speech, together with the areas occupied by alien peoples who in the past have been, for good or ill, strongly associated with the German peoples. Its theme is also evident in the work of another Penckian pupil, Emil Meynen, in *Deutschland und das Deutsches Reich* published in 1935. The theme formed the background to the second world conflict. The concept of *lebensraum* probably the most important geographical concept, was formulated by Ratzel, though it was much distorted by some of his geopolitical successors among the Nazis. In the proper scholastic geographic sense it was inevitably worked over by geographers, and in no small measure by Albrecht Penck.

Penck always took an active part in cartography. His greatest achievement, that moved very slowly and met with much opposition, was the world map on the I/M scale. He actively encouraged the production of topographic maps on a scale of 1:100,000 for the use not only of the military authorities as was the predominant concern but also for the use of scholars and others who were concerned with the study and interpretation of the landscape.

Penck turned to questions of the forms of the surface of the earth throughout the 1890's. He researched on periodicity in the valley development (1884), the formation of through-valleys (1888) and other studies on erosion and denudation. These were based on numerous field studies in Norway (1892), Great Britain (1883, 1884, 1897), Pyrenees (1884, 1897), Canada (1898), Balkans (1900), Australia (1900). Out of all these studies emerged his great book in 1894 on 'the morphology of the earth's surface' which is described thus by H. Louis: '*Es war wiederum eine Grosstat in der Geschichte der erdkundlichen Wissenschaften*'.

In 1905 he wrote of 'physiography'. What he meant by it is clear in his morphology of 1894 and clearer still in a brief statement on 'the new geography' in 1928. In this he declared that the unity of geography should be based on the spatial arrangement of phenomena on the earth's surface in respect of both 'physical' and 'human' processes in so far as such processes help to characterize the unique areas of the earth's surface. Herbert Louis quotes a further statement on this theme written by Penck in a letter one year before his death. The map brings varied phenomena and the landscape into juxtaposition. Hence his emphasis on the use and interpretation and preparation of the topographical map as a geographical fundamental. But all that is important is not on the map. Geography cannot confine itself to maps. One must go out on the land and observe for oneself, to see, for example, crops and architectural styles.

'I have pursued Geography as the study of the earth's surface, in the sense of Ritter and Richthofen's inaugural address. In the beginning I started from minute topographical detail and worked outwards inductively to morphological generalisation. During the First World War, I turned to the vegetation cover of the earth, and then later to politico-geographic question.' In his later years (he continues in the same letter) he turned to 'biological questions', concerned with man on the earth—the earth evaluated as the human habitat. There are several geographies, he writes, one being concerned with the earth as a unit, the other with the areas and the coincidence of the varied aspects.

The peak of Penck's career was the centennial celebration of the *Gesellschaft für Erdkunde*, and the meetings of the Oceanographic Conference, for on both of these, at the age of seventy, he presided. 'Damals war er tatsächlich der uberragende Vertreter der deutschen Geographie.' For the year 1917–18 he became the Rector of the University. After the collapse in 1918, he participated in the movement to revive the spirit of the German people. He was a prime mover in the foundation of the Berlin *Volkshochschule*, and A. Merz, one of his students, became its first principal. Shortly after, he became a member of the Berlin *Mittwochsgesellschaft* and the *Montags Klub*. His inaugural address at Berlin in 1906 was entitled *Beobachtung als Grundlage der Geographie*. Observation in the field is the key note to his scientific work and in his procedure in training a geographer.

One finds a shift in focus in Penck's approach to geography in the forty years of his professional work. Penck's early book on *Das Deutsche Reich* appeared in 1887 in the A. Kirchhoff series of *Länderkunde*. It reveals the strong influence of Richthofen. It is true that August Grisebach brought out his book on the vegetation of the earth in 1871, Ratzel the first volume of his anthropogeography in 1882, and Julius Hann his great work on climatology in 1883, but there remained to be worked out both methods and objectives in the 'synthetic' study of particular areas known in German as *Länderkunde*. For Penck all groups of phenomena had to be interpreted in terms of development and process. For this reason, his treatment of man and the forms of the earth's surface were presented genetically, that is by sequence in time. Penck used for the first time such physiographic terms as *Alpine Vorland* and *mitteldeutsche Gebirgschwelle*. He wrote with sureness and thoroughness on the geological foundations of Germany and its physical divisions, and this is without question a most important feature of his book on Germany. This early work reveals clearly the impact of Richthofen's approach to geography.

Penck's attention to 'regional geography' was not renewed in print until after the First World War. His approach and problems had changed. Louis writes that he was concerned in his later years in the search for the significance of spatial relationships, for the establishment of core and peripheral areas and in the repercussions of natural phenomena on economy, settlement, and culture. Thus, for example, he wrote of *Zwischeneuropa* (1916) as a major trans-continental zone of change in the build of Europe; of the *Grossgau* in the heart of Deutschland (1921); and of the concept of *Deutschland* as a geographic form (1926 and 1928).

Much of Penck's work in the latter half of his professional life in Berlin thus was focussed on the areal associations of societies in their relations to the land they occupied and the habitat they made. Much of his thinking was focussed on the concept of *Lebensraum*, and especially the nature and extent of the German lands. The concepts of *Reichboden*, *Sprachboden*, *Volksboden*, and *Kulturboden* were all developed by Penck, and in these ideas we feel a direct descent from Friedrich Ratzel. Krebs and Meynen in particular followed this thought, but it penetrated to a wider circle and its impact is clearly evident in the field of cultural anthropology and history in the first half of this century.

Penck was not only appreciated by those who worked with and under him, but also he was long recognized as a scholar of inter-national distinction. He received many honours, of which, worthy of mention was his membership of the *Mittwochsgesellschaft*, a very small club of the top members of the University, State and Business. This club was a hundred years old and Penck became its *Kanzler*, although it never had more than sixteen members. Penck was not only a great natural scientist, he was one of the giant makers of modern geography. He was the third in a succession of great geographers at the University of Berlin—Carl Ritter, Ferdinand von Richthofen. These men were among the most distinguished, res-pected, and influential scholars of their time. The above pages are based on the appraisals of his colleagues, and I am also able to quote from his own words. These are contained in a letter written to Carl Troll who made an appraisal of Penck's work as a geographer in 1938 on the occasion of his eightieth birthday. It is particularly revealing and relevant that Penck states that he accepted in the eighties what he calls the *Etagenbau* of Richthofen, as is clearly evident in his book on Germany, but he writes that in his later years, he ceased to accept this and sought for a single integrated focus. This focus he found to be in the unit areas of landscape, and the processes, natural and human, involved in their growth and localization. There is no doubt that this viewpoint, clearly expressed by him in the twenties, was a great stimulus to workers between the wars, among both physical and human geographers.

Since Penck spent the first half of his career at the University of Vienna and the second half in Berlin, it is appropriate to record the sequence of appointments in geography at these centres. The following information is provided in a personal letter from Professor Hans Bobek (see p. 167).

The University of Berlin has had since 1906 one chair in the *Geographisches Institut* and one in the *Institut für Meereskunde* (Institute for Oceanography), the latter founded by Richthofen and separated from the institute of geography when Penck arrived in 1906. The first to hold the chair in the *Institut für Meereskunde* was Alfred Merz, who was one of Penck's students. He died in 1925 and was followed by Alfred Rühl. Rühl became mainly interested in economic geography. He suffered from ill health and died in Switzerland in 1935. Walter Vogel was a second professor (*Extraordinarious*) in the Institute of Geography alongside Penck. After Vogel's retirement in 1938, his chair was converted to a *Lehrstuhl* for the history of Brandenburg, and this was held by an historian, B. Schultze. Norbert Krebs, an Austrian by birth, and a student of Penck in Vienna, was called from Frankfurt in 1927 as the immediate successor of Penck. He died in Berlin in 1947.

The University of Vienna has had a distinguished history of geography also. The first chair was established in 1851 and was held by Friedrich Simony, until his retirement in 1885 (died 1896). Two chairs were established by the University in 1885, one in physical geography, and the other in historical geography. The chair of physical geography was held by Albrecht Penck. In 1906 Penck went to Berlin and was succeeded by Eduard Brückner (born 1862), his junior collaborator in the great work on the Ice Age in the Alps. Brückner remained in this chair until his death in 1927. He was then followed by Fritz Machatschek (1876–1957), another pupil of Penck who remained there until 1934 when he transferred to Munich. The next in succession was Johann Sölch (1883-1952), the last of Penck's Vienna pupils. He was succeeded by Hans Spreitzer, a student of Robert Sieger in Graz who in turn had studied under Penck in Vienna. Spreitzer still holds this post. The second chair in historical geography was first held by Wilhelm Tomascheck (1841-1901) from Graz. He was followed by Eugen Oberhummer in 1903, who came from Munich and held the chair until 1931 and died in 1944. Hugo Hassinger followed, coming from Freiburg in 1931, where he remained until retirement in 1950 (died in 1952). Hassinger trained under Penck at Vienna and for fifty years made notable contributions to the advancement of geography. Since 1951 the chair has been held by Hans Bobek, born 1903 in Klagenfurt, and his main teacher was Johann Sölch at the University of Innsbruck.

Bobek is thus a professional lineal descendent from Penck. Whether by accident or as a result of his training, Bobek has done important work in the distinct fields of glacial geomorphology and human geography. His early monograph on Innsbruck, and his later studies in Persia, are outstanding contributions to modern geography.

It should be emphasized, in conclusion, that while Penck's specialized contributions lay in the area of geomorphology, and especially Pleistocene geology, his main professional work for forty years lay in the field of geography. He made conceptual and substantive contributions to geography, and it is highly significant that many of the distinguished German geographers of the last generation who studied under Penck began as geomorpholigists, but made lasting contributions to the geography of man. This change is evident in the professional growth of Penck himself, in his shift from Pleistocene geology and human pre-history to the imprint of the works of historic man on the landscape.

A feature of Penck's contribution to geography, of which he was probably the greatest founder, needs the strongest emphasis. He originally practised in his book on Germany the mode of presentation of his predecessor, Richthofen. This arrangement he explicitly rejected on many occasions, as is abundantly evident in his research and teaching, in the latter half of his life. The 'new geography', as he called it, must have a distinct and a unique focus, with its own problems and techniques to solve them. This he sought—as he had in his book on the morphology of the earth's surface in 1894, from the standpoint of the 'geomorphologist'—in the landscape unit, that, certainly in Europe, is a combination of physical and man-made elements, and is the result of both physical and social processes. Penck seems later to have thought not merely in terms of the visible landscape scene, but in terms of the variations of regional groupings over the earth's surface. His approach seems to have been much closer to that of Schlüter (and Brunhes) than to Alfred Hettner, who, admittedly in scholarly manner, continued the layered fragmentation first propounded by Richthofen.

Further References

Special attention is drawn to two addresses, which have been freely used in this appraisal, in order to get the assessment of German colleagues, who were closely associated with Penck, in addition to my own familiarity with Penck's works and my frequent discussions with German colleagues. Professor Herbert Louis delivered an address evaluating Penck's work on the occasion of his hundredth birthday (25th September, 1958). This is published by the *Gesellschaft zu Erdkunde zu Berlin* in *Die Erde*, 89 (1958), Heft 3–4, pp. 161–84. Professor Carl Troll delivered an address to the same body in 1938 in honour of Penck's eightieth birthday. This was not published, but Professor Troll has kindly made available to me a copy of it, together with a long and appreciative personal letter from Penck himself. See also Johann Sölch in *Wiener Geographische Studien*, 17 (1948); Norbert Krebs in *Jahrbuch der Deutschen Akademie der Wissenschaften zu Berlin*, 1946–9, pp. 202–12; Edgar Lehmann, *Deutsche Akademie der Wissenschaften zu Berlin*, 64 (1959); G. Engelmann, Bibliographie Albrecht Penck, *Wissenschaftliche Veröff. d. Deutschen Institut für Länderkunde*, 1960, pp. 331–447.

9

Alfred Hettner (1859-1941)

LIFE AND WORKS[1]

Alfred Hettner was one of the most influential figures in Germany in the development of geography. He was the only geographer of his generation—and probably the first—who entered University with the intention of becoming a geographer. He began with a short spell at Halle (1877–8), where he first came under the influence of Kirchhoff who gave to Hettner his first clear insight into the field of geography. He moved to Bonn and here under the influence of Theobald Fischer, turned to his first research on the climate of Chile. He then moved to Strasbourg, where he took his doctorate in 1881 under Gerland, by continuing climatic work on Chile and western Patagonia (published in 1881). He next turned to the geology of the highlands in Saxony. He also became deeply interested in philosophy while at Strasbourg, a fact that explains his essays on the framework and methods of geography which he considered to be his most important work. He returned to Bonn in November of 1881 to study under Richthofen. He found his way to Bogota in Columbia in 1882 as a private tutor to the British Ambassador. Out of this came a book on *Travels in the Columbian Andes* in 1888 as well as various scientific articles. He returned to Germany in 1884 and for the next four years lived in Dresden and Leipzig. He attended Richthofen's *Kolloquium* at Leipzig and continued his work on the geomorphology of the Saxon highlands. He submitted this research for the *habilitation* at Leipzig under Ratzel, who had just succeeded Richthofen on the latter's transfer to Berlin (1887). In the summer of 1888 he returned to extensive travels in South America (with the assistance of A. Bastian, the famous German ethnologist and director of the *Museum für Völkerkunde*). He travelled by boat, rail, stage coach and mule, lived in the open and suffered hardship and sickness.

[1] For this brief biographical appraisal I have drawn mainly from the references given at the end of the chapter.

112

Alfred Hettner

He contracted atrophy of the leg muscles that affected his ability to walk for the rest of his life. These travels covered more than one year and their results are revealed mainly in the letters he wrote to Richthofen, and in his diary, which was subsequently at the disposal of his students. He thus acquired experience in the field as a geographer by serving an apprenticeship, like Humboldt, in South America. He returned to Leipzig under Ratzel in 1890–1, where he gave a course on South America, and where he remained until 1897. He was, on his own statement, not influenced by Ratzel's approach to the geography of man, and laid greater emphasis than Ratzel on the physical basis in geography. In May 1894 he became an assistant at Leipzig, then moved briefly to Tübingen, and finally was called to Heidelberg in 1899. He did not return to any field work in South America, but travelled widely in Russia (1897), North Africa (1911) and Asia (1913–14), accompanied, on several occasions, by Heinrich Schmitthenner. He remained in Heidelberg until his retirement in 1928. (He was succeeded by J. Sölch, then by W. Pantzer, and finally by G. Pfeifer.)

Hettner had a profound influence on the development of geography. Out of some thirty men who took their doctorates under him eleven became professional geographers. Among these are Fritz Jaeger (first professor of colonial geography at Berlin, and then at Basel), Franz Thorbeke (first at Mannheim and then at Cologne), Leo Waibel (Bonn), Oskar Schmieder (Cordoba, Berkeley, Halle, Kiel), Heinrich Schmitthenner (Leipzig, Marburg), Friedrich Metz (Innsbruck, Erlangen, Freiburg), Wilhelm Credner (Munich), and Albert Kolb (Hamburg). These were all geographers, but others attained chairs in the fields of geology in particular. Those who took their *habilitation* with Hettner were Jaeger, Thorbeke, Schmitthenner, and Schmieder.

Hettner and his wife, without a family of their own, had very close contacts with their students. Hettner writes as follows:

I developed a close personal relationship to many of my students, perhaps since I had no family of my own. I name Carl Uhlig, who, though not one of my own students in the strict sense (he took his doctorate under Neumann at Freiburg), was my assistant for the *Geographische Zeitschrift*; Franz Thorbeke, Fritz Jaeger, Leo Waibel, Heinrich Schmitthenner, Ernst Michel, Ernst Wahle, Oskar Schmieder. Among younger men were Wilhelm Credner, Otto Berninger, Johannes Oehme, Martin Rudoph, Paul Gauss, Schwalm, and Albert Kolb. Schmitthenner came to me as my private assistant shortly after the death of his father, who was a close friend of mine, and after obtaining his doctorate. He accompanied me on several extensive travels, and became as a son to me. I believe in fact, that (my students) belong to the best (scholars) of their generation, as in an earlier generation the Vienna students of Penck, and, before him, the students of Richthofen.[1]

Hettner did more than any other geographer of his time to establish geography on a firm philosophical and scientific basis. In 1895 there appeared the first number of his own periodical, the *Geographische Zeitschrift*. Hettner tells us that he had the idea of such a journal while in Leipzig and discussed it with Ratzel and Richthofen. Indeed, he sought to model it on the famous periodical *Ausland*, with which for many years Peschel and then Ratzel were so closely concerned. Hettner also wished the new journal to concentrate on the geographic aspects of current political and economic problems. The first number contained the influential discussion by Richthofen of the treaty of Shimonoseki, and during World War I there appeared many articles on matters pertaining to the wartime situation. Hettner sought always to keep these articles on a scientific plane, unlike the *Zeitschrift für Geopolitik*, that became in large measure an expression of national policy, especially under Haushofer's direction in the thirties. For forty years he used this periodical as a medium for the propagation of his ideas of the scope and method of geography and for his interpretation of current political problems. In 1935 he passed on the editorship to Heinrich Schmitthenner and publication continued until 1943 when the journal was suppressed.

It is highly significant that publication was resumed in January of 1963 with Gottfried Pfeifer as the principal editor. Pfeifer writes in the first number of the revived journal:

It is natural that two decades have not passed without repercussions on geographical science. Now, as before, it is necessary not only to advance detailed research in the field of our science, but also to formulate the idea of its unity. The present position of geography makes this objective especially urgent. We are in a period of active development in almost all areas. Classical geographical climatology is threatened to be divided and uprooted by meteorology. Geomorphology is in danger on one side from climatology, on the other side from sedimentary petrography, and is tending to split in such a way that it is departing from the geographer

[1] This is taken from *Alfred Hettner: Gedenkschrift zum* 100 *Geburtstag, Heidelberger Geographische Arbeiten*, Heft 6 (1960). See below.

though he needs it as the basis of his regional geography. Anthropogeography is being invaded by the methods and goals of statistics, history, archaeology and sociology. Entirely new fields, such as social geography, ranked as central themes in the last German geographical meetings. This interlocking of disciplines in border problems is a feature of our time. On the other hand, it is essential for geography continuously to explore the territory in the border areas of other sciences. Many of the branch fields to which it has given birth are now independent, but pay little attention to their relation to others. This is also true of publications and numerous publications of key importance have appeared in periodicals of other disciplines. The purpose of the new *Geographische Zeitschrift* will be to fill this gap. It gives clear expression to the development of geography under stimulus from Hettner over a period of more than fifty years and the new journal will continue the objectives laid down by its founder.

In 1891 Hettner was approached by a publisher (on the recommendation of Penck) to prepare the text for a new *Handatlas*, to be modelled on the recent atlas prepared by F. Schrader in France. He undertook this task but had to forego the publication of his field work on Peru, later handing over his route maps and notes to Carl Troll. Out of this work grew his large volume on the regional geography of Europe, published in 1907, and its much delayed continuation on extra-European areas in 1924. These two books carry the title of 'foundations of regional geography' (*Grundzüge der Länderkunde*) and were revised in the twenties. He later published *Comparative Regional Geography* (*Vergleichende Länderkunde*) in four volumes (1933–5). His other important books were on *Russia* (1905), *England's world domination and the War* (1915), *Surface Forms of the Continents* (1921 and 1928), and *The Spread of Culture over the Earth* (1928 and 1929). Hettner initiated the idea of a series of volumes on the regional geography of the world entitled *Handbuch der Geographischen Wissenschaft*. This was finally carried out by a younger colleague, H. Klute, though some of the contributors were originally chosen by Hettner. It falls into eleven volumes and was completed in 1940. It is one of the monumental works of geographic literature. Some of Hettner's material was published posthumously as *A Geography of Man*. This is in three volumes— the *Basis of the Geography of Man*, edited by H. Schmitthenner (1947), *Transport Geography*, also edited by him (1952), and *Economic Geography*, edited by E. Plewe. Hettner's essays on the theory of geography, published between 1895 and 1905, plus additional sections, were published in a single book in 1927, with the title *Geography: Its history, character and methods* (*Die Geographie Ihre Geschichte, Ihr Wesen und Ihre Methoden*, Breslau, 1927). The following comments are taken from this source.

The historical development of geography reveals two concepts of its nature. According to one view geography is a general science of the earth, whereby general geography is basic and special geography (the study of particular areas) is secondary. According to the other view, geography is the study of the earth's surface, and areal character (*Länderkunde*) stands in the foreground and general comparative regional geography is secondary. As long as the system of sciences was based on distinctions of material investigated, geography could stand as an earth science. However, Hettner maintained that this is an erroneous interpretation, since the chronological or historical and chorological or areal approach stand in clear distinction alongside the systematic study of particular sets of data, so that a chorological science of the earth's surface is not only justified, but it is essential to a complete system of sciences.

Geography has often been regarded as the science of the earth as a unit—hence its German alternative name *Erdkunde*. In fact, the study of the earth as a whole falls to several sciences that have developed recently. In a sense, geography has given birth to them and they now stand independently. Such are geophysics, that early in the nineteenth century was referred to as physical geography; geochemistry (minerology, petrography and soil science); astronomy and mathematical geography; and aspects of plant and animal study. The historical aspects are closely related to historical geology rather than to their variations on the earth's crust.

Geography is the chorological science of the earth's surface. This characterized the work of Strabo, except that he defined his approach to the study of countries as chorography, and the same idea is basic to Ritter's concept. The geographical sciences deal above all else with the areas of the earth's surface, in so far as these have material content (*irdisch erfüllt*), that is, with the description and spatial relations of places. They are distinguished from the historical sciences, that investigate and present the sequence of events or the succession and development of things.

Hettner maintained that geography is not the general science of the earth (*Erdkunde*), but it is also ambiguous to regard it, as Richthofen did, as the science of the earth's surface (*Erdoberflächen kunde*). Its traditional and logical field, he argued, is the earth's surface according to its localized differences—continents, lands, districts and localities (*Erdteile, Länder, Landschaften und Ortlich-*

116

keiten). It seeks to define and describe these unit areas (*Länderkunde*), and to compare them on an inductive world-wide basis (*vergleichende Länderkunde*). This means that its concern is with terrestrial things in terms of their spatial arrangement, in contradistinction to the systematic sciences, that are based on the separate categories of things, and the historical sciences based on sequences with respect to separate categories of things or sequences of human events. The study of the differential associations of phenomena over the earth is the keynote to Hettner's concept of geography.

This conceptual approach must guard against the idea of geography as the study of distributions. It is not necessarily concerned with the distributions of particular phenomena, which are of more concern to the systematic sciences—the distribution of individual plants or animals, or minerals. The geographic approach enters only when such a distribution is an element of an areal association of phenomena which gives character to an area on the earth's surface—*die irdische Erfüllung der Erdraüme* of Ritter. Wallace made this distinction clear when he spoke of the distribution of individual species or orders as geographical zoology, but the science of the study of the associations of fauna in particular areas as zoological geography. The same holds true for the study of plants, minerals, landforms and the atmosphere. Concern with the distribution of individual phenomena and their causes is the study of the systematic sciences. Geography begins with the spatial associations of phenomena that give character to areas of different orders. Similarly, the distribution of a weapon or a custom belongs to the realm of ethnography. The distribution of a particular fact of economics can only be regarded as geographical economics, but economic geography on the other hand, deals with the localized economic phenomena that are areally associated and give character to areas. Man, says Hettner, cannot be regarded as the focus of such study, as was regarded by Ritter and his successors, because the idea of Nature v. Man and the exclusive concern with man-land relationships, with an interdependence of (so-called) Geography and History is logically incompatable with the chorological approach. The six realms of the physical world—land, water and air, plants, animals and man—must each be studied on their own merits and in their spatial interconnections and causal relations in so far as they are associated in area. It is true that Man, as a primary agent of change in transforming the earth, will play a dominant role in this study, but this is not the measure of the objectives of the study. The essential concern is with the phenomena on the earth and their arrangement as associations in area.

Ritter placed man in the focus of geographic study, but Hettner asserted that Man must be considered alongside the other realms of Nature, and that the view of Man's dependence on the earth's surface which was adhered to by Richthofen in his China work and also by W. M. Davis, contradicts the traditional concern of geography with the total character of areas. 'It is neither a natural nor a human science . . . but both together' for it is concerned with the 'distinctive character of lands', in which Man and his works are of greater or lesser prominence. In this sense the study of lands—*Länderkunde*—unites the physical and human aspects and there is no dualism.

He claimed that Ratzel regarded geography as a distributional science and stressed *space* rather than *land*. Götz, the founder of economic geography in 1882, followed Ratzel in regarding the geography of trade (*Verkehrsgeographie*) as a science of distance so that its central theme is the conquest of distance. One assumes space without the reality of content. Hettner also attacked the 'landscape' concept of Schlüter and Bruhnes (see Chapters 10 and 17), as aesthetic geography. Geography cannot be limited to the colour of soil, without reference to its physical and chemical qualities; it cannot limit climate to the observation of the colour of the skies and the form of the clouds. In reality the 'landscapists' do not exclude these things, but bring them in by the back-door. A sharp distinction between the visible and the non-visible is just not realistic, and indeed this approach would exclude traditional fields of geography such as political and economic geography and 'ethnogeography'. Both Schlüter and Bruhnes recognized this, he said. (Hettner interpreted the views of these two men too literally and narrowly.)

He continued that geography as a chorological science cannot find unity in the appearance (*Bild*) of the landscape, but in the inner character (*Wesen*) of the lands, districts and localities of the earth's surface. This end is attained in two ways. First, by the recognition of geographical complexes and systems, e.g. river systems, atmospheric circulations, trade areas, etc. for no spatial system can be understood alone, but must be related to others. The second consideration is the causal interdependence (*Zusammenhang*) of various sets of spatially arranged phenomena. Individual distributions alone are of no geographic significance in characterizing an area, but a single distribution is of significance in characterizing an area in so far as it is coincident with other areal phenomena with the same distribution—i.e. in so far as it is geographically efficacious.

Geographical study does not follow the passage of time, although this methodological rule is often overlooked. It takes a limited cross-section of reality at a particular time and uses development through time only for the explanation of the condition in the chosen period. 'It needs a genetic approach, but not to become history.' The geography of past periods would seem to lie mainly in the competence of historians and archaeologists, just as palaeogeography lies in the hands of geologists. This view of Hettner's leads to discussion of the controversy between German geographers, led by him and Passarge, and W. M. Davis, the geologist, who was the primary founder of geography in America around 1900, and who had a great impact on scholars in Europe in the opening years of this century.

THE DAVIS–HETTNER CONTROVERSY

The so-called Davisian system of 'explanatory description of land forms' was the subject of bitter criticism from certain German geographers in the years following his lectures in Berlin. These lectures were given in 1908 at the invitation of Albrecht Penck, and translated into German by one of his colleagues, Alfred Rühl, with the title *The Explanatory Description of Land-forms* (*Die erklärende Beschreibung der Landformen*), published in Leipzig in 1912. The work was never translated into English though the ideas are to be found in many of the articles published both before and after this time by Davis. Alfred Hettner's criticisms first appeared in the *Geographische Zeitschrift* in 1911, but his major presentation was *Die Oberflächenformen des Festlandes* in 1921 (second edition 1928).[1] About the same time Walther Penck's posthumous work on *Die Morphologische Analyse* appeared, edited by his father. W. Penck was concerned with a scientific analysis of the development of landforms in terms of the development of slopes as a contribution, he claimed, to historical geology. Penck's contribution was subjected to a searching analysis in comparison with the Davisian system in a symposium at the meeting of the Association of American Geographers in 1940.[2] This invaluable presentation by leading American geomorphologists is available to the interested reader and its conclusions are summarized elsewhere. It is much more relevant here to examine the attitude of the German geographers to the Davisian system, since this has nowhere been adequately presented in English. The attacks on both sides were bitter and were quite clearly based on

[1] Siegfried Passarge also was involved in this controversy. He made his rebuttal to the deductive method of Davis in his book *Physiologische Morphologie*, published as an article in *Petermann's Mitteilungen* in 1912. See p. 138.

[2] *Annals of the Association of American Geographers*, 30 (1940), pp. 219–79.

different ideas of the problems, purpose and methods of geography, as well as on differences of national temperament.

It must be recalled that in the first two decades of this century, there were a number of mature geographers in Germany, scholars of wide repute, whose ideas were far in advance of their contemporaries in Britain or America, in both field work in widely scattered areas of the earth and in formulation of geography as a distinctive discipline (p. 59). Moreover, both Richthofen and Albrecht Penck had already produced in the 1890's the first basic works on land-forms, their classification, origin and development. All these men were concerned, as geographers, with description and association of phenomena on the earth's surface. They were not nearly so fettered by environmental determinism at this time as were Davis in America and Herbertson in Britain. They sought to organize geography as a discipline which they regarded as being primarily descriptive and empirical. They considered that the explanation of distinct land-forms as well as of man-made (culture) forms imprinted on the land is to be found in the result of a variety of forces operating in different ways in different areas. Davis presented in 1903 a method or 'scheme' of explanation of land-forms, which in the understanding of the detailed factual articulation of the earth omitted all that was irrelevant to the system. The Davisian concept of the geographical cycle (so-called) and its operation on a land surface was envisaged in terms of 'structure, process and stage', and the land surface had features that were theoretically associated with 'youth, maturity and old age'. It was argued by Hettner that this was an arbitrary scheme, that it neglected, even ignored, other contributory local factors operating in an area, such as weathering, variations in structure and lithology, and climatic factors. These factors were not only factors of climatic change during the cycle of erosion, but also differences in the combination of the processes of erosion and deposition in different milieus of the world. Davis' 'normal' cycle was developed and primarily applied to, the humid areas in Europe and America. German scholars had already undertaken extensive scientific observations in arid zones of the earth, and they claimed that the development of land-forms through the operation of the wind and rain and weathering, and the resultant processes of erosion and deposition, vary in different climates and with changes in climate in one and the same area. The soil and vegetation cover have repercussions on surface changes. Davis' scheme was arrived at inductively from case studies, by Davis and two of his older contemporaries, Powell and Gilbert. These case studies, as well as those of other workers in Europe in the sixties, found recognition and praise from both Hettner and Passarge, but they both rejected a generalized system that was deductively applied so as to portray, on this limited explanatory base, the surface features of an area. They argued that the geographic problem was to characterize land-forms in their locally repetitive forms and to explain them in terms of *all* the forces operating on them. Davis did not study the land-forms as they occurred in reality, but defined and explained them deductively within the framework of a theoretical system, which, it was argued, itself contained fundamental errors. The system was designed, as Davis very cogently insisted, to assist the memory and to give a single key of cause and effect in understanding landforms.

It is to be expected that these arguments of Hettner, that reflect the viewpoint of the geographers and their objections to the Davisian system, would meet with rebuttals from the States. They received attention in particular from both Davis and Bowman. Davis made scathing and personal comments in a review of Hettner's work.[1] This, he described, as full of 'homilies, truisms, hesitations, obstructive misunderstandings, and disputatious objections'. He claimed that the book was a 'concientious, but reactionary protest against what its author regards as the too hasty methods of certain other geographers'. He described the book as 'a diatribe against the erosion cycle'. Davis was evidently, and understandably, angry, but he really did not (at that time and on this occasion) understand the point of view that Hettner was trying to get across, the viewpoint of a geographer rather than that of a geologist.

Both Davis and Bowman became further involved in trying to bridge the gap, which was in fact a gap between geography and geology. Bowman concluded that the difference arose from 'an unwillingness to accept a terminology of foreign origin and in part from an inextricably persistent misunderstanding of *stage* to mean *age*, a failure to see that the word stage is employed as a measure of development, not of time'.[2] This may have been correct, but it is not the essential issue of the geographical argument. Bowman then objects, as did Davis in his review of Passarge's *Die Grundlagen*, to the 'overdoing of empirical description' as opposed to the genetic approach of the geographical cycle' (this term being used, by the way, as opposed to the erosion cycle). This approach undoubtedly offended the professional susceptibilities of Hettner as a geographer, for, says Bowman, empirical description, especially in the form of

[1] *Geog. Rev.*, 13 (1923), pp. 318–20.
[2] I. Bowman, 'The Analysis of Land-forms: Walter Penck on the Topographic Cycle', *Geog. Rev.*, 16 (1926), pp. 122–32.

interminable lists of classified landscape phenomena[1] (which makes up the bulk of the first volume of Passarge's work) offers to a student 'an ideal of complete meaninglessness and deadly monotony'. Bowman further argues, '. . . Ideas run the world, not outlines or catalogues'. We understand and sympathize with this brilliant aphorism, but we feel now that his reaction reveals a lack of real understanding of what Hettner (and Passarge) were driving at *as geographers*. The argument seems to be that empirical classification and mapping *per se* is deadly dull as an end in itself. Such was never the end of either Hettner nor Passarge, but an essential means to an end.

DAS LÄNDERKUNDLICHE SCHEMA

Geographers around 1900 continued to emphasize the Physical Environment *versus* Man as a central theme. As a result one could read studies about particular areas without having the slightest idea of the appearance and arrangement of the landscapes as the composite man-made habitat built on the basis of terrains. Indeed, the mode of geographic study of areas or countries tended to follow a routine, in which various categories of facts were examined in their geographical distribution, arranged in a more or less stereotyped order *ad seriatim:* position, geology, relief, climate, natural resources, prehistory, medieval period, distribution of population, occupations. routes, and political divisions. This arrangement was based on the assumption that these followed each other in a sequence of cause and effect. Most of the material was drawn from other sources and the geographer contributed nothing new to it, apart from drawing the maps, and showing, above all else, the relationship of these phenomena to the 'physical base'. It is quite conceivable however, that in any one area the six realms of Richthofen and Hettner (land, water, air, plants, animals, man) could be studied independently, and thoroughly investigated each in itself, although the interconnections between them (as opposed to their connections with the terrain and position) were minimised. This became a standard method of treatment in Germany, as it was also in the French monographs in the 1900's. It was referred to in Germany as the *länderkundliche Schema*,[2] and it served as a frame for many scholarly German works

[1] Vol. 1 of Passarge's work was apparently the target of this criticism. The remaining three volumes were almost ignored by Bowman and apparently not reviewed, for they contained the answers to the criticism. Bowman was too previous in his condemnation. See p. 137.

[2] A. Hettner, 'Das Länderkundliche Schema', *Geographische Zeitschrift*, 39 (1933), pp. 93–8, and E. von Drygalski in *Pet. Mitt.*, 78 (1932), pp. 6–7.

from the 'eighties to the 'twenties. This approach, for example, will be found at a high level of scholarship in Philippson's *Das Mittel-meergebiet*, first published in 1904 and dedicated to Richthofen; in Gradmann's two volumes on *Süd-deutshland* (1931); and Hettner's *Europa*, in *Grundzüge der Länderkunde*, that emerged in 1907 (second ed., 1925) from a text written in the 1890's to accompany an atlas. Though this schematic approach has long been discredited as a stereotype (notably by Penck in the twenties, who, following Richt-hofen in the 1880's, accepted it), it still appears in certain recent so-called 'regional geographies' in Britain and America.

In Germany, the main stimulus to the study of small areas, was due to a central commission on the regional geography of Germany established in 1882 by F. Ratzel. This committee sponsored the publication of studies in a series called *Forschungen zur deutschen Landes—und Volkskunde*. The early publications lacked the 'synthetic method' that was advocated at that time by Vidal de la Blache in France. They covered more specialized topics, such as 'The towns of the North German Plain in relation to the configuration of the ground' by E. Hahn; 'The plain of the Upper Rhine and its neighbouring highlands' by R. Lepsius; 'Mountain structure and surface configuration of Saxon Switzerland' by A. Hettner; 'The Erzgebirge: an orometric-anthropogeographical study' by J. Burgkhardt. This publication continues to this day and is the longest series of geographical studies in print. Its name was changed in 1946 to *Forschungen zur deutschen Landeskunde*, whereby the traditional tie with ethnography was broken. Over a hundred volumes have been published since 1946. Almost all of these deal with some special topic of physical geography or anthropogeography either for a small section of Germany or for the country as a whole.

Reference should be made to early studies in *Länderkunde* that appeared in Germany before 1900 and were based upon first-hand field observations in remote areas of the world. Among these were A. Supan's physical geography of Guatemala (1895), K. Hassert's work on Montenegro (1895), J. Partsch's study of Corfu (1887), A. Philippson's work on the Peloponnesus (1891) and above all, A. Hettner's study of the Cordillera of Bogota (1892).[1] The last study is accompanied by four maps, all based on data collected on traverses in the field. These show; geology; relief (*tierra caliente, templada, fria, and paramo*); places by size; and density of population, showing (1) uninhabited, (2) very thinly inhabited, (3) settlement oases, and (4) four levels of density, 10–25, 25–50 and 100–150 per square kilometre.

One of the outstanding senior students of Hettner, namely

[1] *Pet. Mitt., Erganzungsband*, No. 104, 1892.

Friedrich Metz, has recently written[1] that for Hettner *Länderkunde* was the 'crown of geographic work', yet 'it is particularly surprising that scarcely any of the thirty doctorate dissertations prepared under him at Heidelberg handled such a theme'. Many were concerned with land-forms in an area accessible to field observations from Heidelberg—Jaeger on the Odenwald, Schmitthenner on the northern Black Forest and Credner on the crystalline sections of the Odenwald and Spessart. Thorbecke worked on the climate of the Mediterranean, and Waibel on the animal life of Africa and its relations to the economies and modes of life of human societies. Other studies were concerned with aspects of settlements or economies in particular areas of the southwest of Germany, such as Metz' own work on the Kraichgau. Much of this work was presented to and discussed at a seminar which never had more than about 25 participants and, says Metz, contained such a diversity of problems and approaches that one could seriously doubt whether there was such a thing as a Hettnerian school of geography, for Hettner exercised no kind of specific constraint on his students.

It would appear from this list that Hettner's approach to the training of a geographer was based on field observation in an accessible area, and the selection of a particular single geographic aspect of land or people as the focus of investigation, be it land, climate, settlement, economies or social life. This segmented approach is abundantly evident in Hettner's own work. Attention has been drawn to this in the presentation of his field work in Columbia in the early 1890's, and it is found in the presentation of his text books two and three decades later. All this is strikingly different from the Vidal school. This difference was due, in part, to the much more exacting academic demands on the French doctorate than in German Universities. But it is also due to the difference of conceptual approach between Hettner and Vidal de la Blache, who insisted on research into the ways in which spatial phenomena of land and people are interdigitated. This is made quite clear in the changing form of the exhaustive regional monographs that have now been published for nearly seventy years as the essential entrée to the geographic profession in France.

Further References

The most authoritative and recent source is the memorial volume to

[1] F. Metz, in 'Alfred Hettner als akadamischer Lehrer', *Gedenkschrift zum 100 Geburtstag, Heidelberger Geographische Arbeiten*, Heft 6 (1960), pp. 34–5.

Alfred Hettner

Alfred Hettner—*Alfred Hettner: Gedenkscrift zum 100 Geburtstag, Heidelberger Geographische Arbeiten.* This contains two essays by Ernst Plewe and Friedrich Metz on Hettner's work as a researcher and a teacher, invaluable autobiographical sketches, edited by Plewe, and a full bibliography. There is also a long article on Hettner, written by Heinrich Schmitthenner, who knew him professionally and personally better than any other person, in *Allgemeine Geographie des Menschen, Band I, Grundlegung der Geographie des Menschen*, edited by Schmitthenner, Stuttgart, 1947, pp. 11–43. Reference should also be made to R. Hartshorne's *The Nature of Geography* particularly pages 137–44. None of the writings of Hettner himself are available in English—hence the great importance of Hartshorne's evaluation—but they must be read in order to understand the purpose and method of his work. This applies especially to his classic work *Die Geographie, Ihre Geschichte, Ihr Wesen und Ihre Methoden*, Breslau, 1927.

10

Otto Schlüter (1872-1952)

In the attempt to organize geography through synthesis after a generation of analysis and dualism, Hettner sought the answer in the areal associations of terrestrial phenomena and the recognition of the content of terrestrial areas. A rebuttal to this view was presented at the turn of the century by Otto Schlüter which has had profound repercussions on the subsequent development of the field of geography. Schlüter raised objections to both the environmental emphasis on land-man relationships of the current schools—Ratzel, Davis, Mill and Vidal de la Blache—and to the view of areal associations of terrestrial phenomena as expounded by Hettner. Neither, it seemed to Schlüter, provided a distinctive field of study.

Otto Schlüter, born in 1872 in Westphalia, began his academic studies in German language and history in the University of Freiburg and Halle and this interest persisted throughout his long life. This falls into three parts. From 1895 (at the age of 23) to 1910 he developed his conceptual framework of human geography. From 1911 to 1951 he was director of the geographical institute in the University of Halle and carried out his major researches on the geography of settlement in central Europe. He retired in 1951 as professor emeritus and then prepared the three volumes of his fundamental work on the geography of settlement in central Europe, that had occupied his attention for the best part of fifty years.

While at the University of Halle as a young student he was greatly impressed (as was Hettner) by the lectures of Kirchhoff, the geographer, and the first professor of geography in that University. It was through Kirchhoff's influence that young Schlüter shifted his interest from Germanistic studies to geography. He pursued and concluded his formal geographical training at Halle with a dissertation on *Siedlungskunde des Thales der Unstrut*. This work followed the traditional pattern, and he subjected it to trenchant self criticism in his *Habilitation* seven years later. This earliest work revealed two traits, however, that affected all his subsequent research—the use of

Otto Schlüter

place names in interpreting modes and sequence of human occupance; and the battle on the Unstrut in 531 which he came to recognize as a cultural break in time and space in his later studies in historical geography in central Europe.

In 1895 Schlüter moved to Berlin. Here, under Richthofen's strong influence, he developed his stand on the scope and method of human geography. In 1899 Schlüter wrote his 'remarks on the geography of settlement', in 1903 his *habilitation* thesis on the settlements of northeastern Thuringia, in 1906 his 'objectives of the geography of Man', and in 1907 'the relation between Man and Nature in Anthropogeography'. At this last date he was only 35 years old. During this first period (1895–1910) he was in Berlin, except for a short time spent in Bonn (where he launched a historical atlas of the Rheinland).

In 1911 Schlüter accepted the chair at the University of Halle in succession to Philippson. Here he remained for the rest of his life. Though he retired in 1938, he remained at the University intermittently until 1951 when it was finally taken over by E. Neef and R. Kaubler. Schlüter was thus at the helm in Halle for forty years. It needs to be emphasized that although his teaching was directed to a scientific study of human settlement, he also required a thorough training in geomorphology, and was well qualified to teach it.

During this latter half of his life he turned from methodological matters to substantive research in central Europe. In the thirties he undertook an atlas of *Mitteldeutschland*. He recognized the significance of *circa* 500 A.D. in the German settlement of central Europe and used place names, prehistoric finds, historical writings (of Tacitus, Strabo and others) to reach further understanding of the landscapes of central Europe at this time (*frühgeschichtlichen Altlandschaften*). In his last years in the fifties he produced three volumes on the settlement areas of central Europe in early historical times (*Siedlungsräume Mitteleuropas in frühgeschichtlicher Zeit*) (1952, 1953 and 1958).

This last named major work is based upon a remarkable map of the forested and forest-free areas of the whole of central Europe at the dawn of the era of medieval forest clearance. By this period Schlüter means, in modification of the precise definition in his early years, the centuries following the Roman domination but before the onset of the clearance of the middle ages. He found difficulty in showing significant periods on one map, and resolved this by using several shades of green to show the forest clearance in their historical sequence. This succeeded in combining purpose and attractiveness, and scholars concerned with the history and geography of Germany

127

receive from this map a fund of information. One does not recognize in the completed map the diligent research into extremely varied sources upon which it is based. The first volume contains an explanation of the method, and the remaining two volumes examine the regional variations. This is one of the great scholastic achievements in the field of historical and geographical analysis.

Otto Schlüter wished to focus geography on the study of the visible landscape. This would exclude the investigation of human life, thought and organization, which, though having marked geographic localization, are surrendered to the other sciences which clearly are concerned with the geographic distribution of their phenomena of study. Such elements and their distribution must be understood, borrowed and used to interpret the landscape that, as an architectural deposit, mirrors the culture and economy of human groups. Thus political facts are only examined in so far as they affect man's distinctive politico-geographic area. As Ratzel wrote, 'similar functions began similar forms', so the elements of the landscape, examined as to their forms, functions and grouping, and their areal associations at various levels, should be, according to Schlüter, the focus of geographic enquiry. This is the scientific study of the surface of the earth, whether the portion of earth happens to be wild, or whether deeply transformed by man, or whether partly— either obviously or problematically—affected by man. Herein lies a refinement of Ratzel's concept of landscape (*Landschaft*), for the distinction is made by Schlüter between cultural landscape (*Kulturlandschaft*) and natural landscape (*Naturlandschaft*). It early became obvious to Schlüter and others (especially those working in Europe) that the distinction between the two is difficult to define or conceive. Landscapes have changed through natural processes through time. Climatic changes for example, affect the kind of vegetation. Man himself, by accident or design, has transformed the vegetation and the surface configuration by burning and felling trees, by drainage, by the introduction of new crops and animals (both domesticated and others brought with him, such as the rabbit in Australia). Thus, in man's occupance of the earth one has to take a period of time at which one can assert with some certainty that before man made his major impact on the land the vegetation had certain predominant features. Schlüter, therefore, spoke of the *wild landscape* (*Urlandschaft*) in western Europe, and took it to mean the landscape of woods and clearings such as existed at the dawn of modern man's great transformation of it around 500 A.D. The study of landscape is then a matter not only of classifying the categories of phenomena and determining their distribution and associations, but of examining their characteristics through the process of change through time. Also there arises the problem of classifying and mapping terrestrial areas on the basis of the degree of impact of man upon the landscape, and areal and world-wide studies of this kind have been carried out. Of particular importance in this connection was E. Friedrich's study of 'robber economy' and its various forms and degrees of impact, that has served as a basis for evaluating and classifying the cultural landscape. Schlüter's outstanding work, as already noted, has been on the distribution of woodlands and clearings in central Europe around the middle of the first millennium A.D. which was the effective beginning of extensive transformation of the modern landscape, and whose characteristics can be deduced from documentary records, botanical evidence, pollen analysis, etc.

Schlüter's main case for focussing the geography of man on the interpretation of his work in the landscape was made in his paper on *Ziele der Geographie des Menschen*, Munich, in 1906. 'What we desire is . . . limitation in the subject matter and objectivity in the observation'. He suggested that human geography should aim at 'the recognition of the form and arrangement of the earth-bound phenomena as far as they are perceptible to the senses'. He claimed that his method was morphological and its procedure parallel to the study of land-forms. All human distributions of non-material character, such as social, economic, racial, psychological and political conditions, are excluded from this study as ends in themselves and are considered only in so far as they are contributory to an understanding of the evolution and character of the landscape—a mode of approach that is parallel to that of physiography, that is (or should be) concerned *not* with the exclusive study of the processes— e.g. mechanics of erosion—but with the facial expression of these processes on the landscape. The cultural landscape embraces, according to Schlüter at this date, both immobile and mobile forms. The immobile forms need to be explained in terms of 'all the effects of which every period and every culture, according to the measure of its forces, has wrought upon the landscape'. The mobile forms include Man, together with his works, his movements, and the distribution of each and all. The cultural landscape includes, therefore, not only the routes and the route patterns, but also the men and the goods which move along them.

It follows from Schlüter's concept that a small unit of areal association is a physiognomic unit in which all the 'perceptible' phenomena, both natural and human, which have areal significance, form together a distinct association, and that, as far as man is concerned, the association is rooted in similarities of function or organization or common origin. It is fundamental to this concept that the first aim in treating an area should be to describe accurately the landscape on a disciplined basis of classification of forms. This aim, as Schlüter repeatedly stated, is frequently sacrificed in the search for causal relations. Secondly, areas must be examined as to the origins, functions and grouping of landscape elements. This approach therefore demands the study of the cultural landscape as it has evolved from the original or natural landscape. The areal groupings of landscape elements in the existing landscape (which is the primary goal of the study) are to be explained in terms of the way in which they are tied together in the present functional organization of an area (what Richthofen described as the dynamic relationships of areal functional organization) and in terms of their genetic development, through which past survivals in the landscape may find their explanation (e.g. the precise extent of a three-field system and the type of settlement associated with it). This morphological approach is clearly a fundamental departure from the Darwinistic environmental approach that so deeply affected the thought of W. M. Davis, H. R. Mill and others at this time. It became widely current in the 1900's—but, incidentally, had virtually no impact in either Britain or America.

The views of Schlüter are logical and give a definitive field of operation, with a distinct field of problems, and a constantly changing situation through the continuous changes of the structure of human societies and the changing nature of their demands on their habitat.

The following statements by outstanding German geographers in the twenties and thirties afford a measure of the minimum of common agreement. Albrecht Penck claimed in the mid-twenties that 'the essence of geography is that it deals with terrestrial areas', and that 'the visible content of the landscape determines the content of modern geography'.[1] He stated that the smallest region (*Landschaft*) is an area which has a unity of Form and Function. Such an area, he called a 'chore', following, in a more limited sense, the term coined by J. Sölch in 1924. Norbert Krebs, the pupil and successor of Penck at Berlin, held the same view. He added that to understand the landscape and the areal relationships of its elements, it is essential to examine the social structure and peculiarities of the society which created and lives in it.[2] Leo Waibel, a pupil of Hettner, and a

---

[1] A. Penck, 'Neuere Geographie', *Jubilaüms—Sonderband der Zeitschrift der Gesellschaft für Erdkunde zu Berlin*, 1928, pp. 31–56.

[2] N. Krebs, 'Natur- und Kulturlandschaft', *Zeit. d. Ges. für Erdkunde zu Berlin*, (1923), p. 81–95.

leading geographer until his death in 1951, declared that 'the study of the cultural landscape, stimulated by Schlüter, stands in the foreground of geographical interest. It is the corner stone of geographic work as the areal expression of man's work in creating his habitat.'[1]

Note here the reference to habitat. The focus has clearly shifted. The 'natural environment' is an abstraction, for it has been radically transformed by Man in creating his habitat. Over the last fifty years attention has shifted from the study of 'natural environment' as the home of man to the study of the human habitat as the creation and workplace of human groups, whose reactions are conditioned by their group habits and heritage. It finds its focus in what Carl Sauer has called habitat and habit, or in what I have alternatively described as landscape and society.[2] It is concerned with the location and grouping of human groups in their diversified habitats. It probes into systems of regional definition and analysis, as well as into the precise nature of man-land relationships that emerge through the adjustment of human groups (in the light of what James calls their attitudes, objectives and technical abilities) to the diversified environments in which they operate.

The work of Otto Schlüter and its significance in the development of German geography were summed up in the mid-fifties by Herman Lautensach.[3] He writes that Schlüter founded 'modern cultural geography' on a base comparable to that of geomorphology. His appraisal may be paraphrased as follows. Schlüter, like Richthofen, drew a distinction between *genetic* analysis, that, is the process of change in form, and *causal* or *dynamic* analysis, that is, the causes that lie behind the process of change. Lautensach recognizes three kinds of causality—mechanical, biological and psychological. This distinction in geographical work, he writes, is recognized by Troll and Bobek. The field covers the visible landscape and the distribution of people—the works of man and the distribution of man.

The regional concept (*Landschaftsbegrift*) of Schlüter, like that of Hettner, is a geographical portion of the earth's surface that stands out from its surroundings. Hettner stressed the *Wesen* or personality of an area as based upon the similarities of contiguous places and their phenomena, spatial cohesion, and the causal cohesion of the

---

[1] Leo Waibel, 'Was verstehen wir unter Landschaftskunde', *Geog. Anzeiger*, 34 (1933), pp. 197–207.

[2] See 'Landscape and Society', *Scottish Geog. Magazine*, 55 (1939), pp. 1–15 for a further discussion.

[3] H. Lautensach, 'Otto Schlüter's Bedeutung für die methodische Entwicklung der Geographie', *Pet. Mitt.*, 96 (1952), pp. 219–31.

various natural realms in this area. Schlüter stresses the *Bild*, the association or assemblage in space of landscape elements as the essence of his landscape unit. The morphology of landscape is associated with the ecology of landscape; it interprets the distribution of language, religion, law, art, technics and state only in so far as they are relevant to the landscape. They are pursued in so far as they are contributory causes, not as ends in themselves.

Hettner was opposed to Schlüter's limitation of geographical study to the visible landscape. He was concerned with the uniqueness of areas, whether this uniqueness was evident in the visible landscape or not. When Penck defined his 'landscape unit' in these terms in 1928, Hettner was again in opposition. He recognized the focal interest of landscape, but refused to recognize the limits set by it on the study of the human facts in space.

Today, German geographers, writes Lautensach, universally recognize the landscape as focal to their field of study. Since the war it has been objected that researches are wandering into peripheral fields and losing a focus. Hettner and Schlüter prevented this in the first half of the century. This focus, concludes Lautensach, must be maintained and this can only be done by a clearer definition of the regional concept and the problems it poses.

Hettner's approach provided an eclectical structure for geography as a distinct discipline. Schlüter's concept of cultural geography—the landscape concept—has stood the test of time and has continued to grow by increasing clarity in concept and practice, and in the shaping of new problems relevant to a rapidly changing society. Leo Waibel, clearly expressed this in his own research work. In 1933 he wrote:[1]

Schlüter was the first to raise the landscape-forming activity of man to a methodological principle. Going out from the application of the concept of the physiognomic build of the landscape, he gave to the geography of man a corporate substance for research, which can be worked out according to the same method as physical geography—the cultural landscape. This can be examined from the standpoints of its morphology, physiology, and developmental history, just as the visible phenomena of Nature in the build of the landscape. Between physical geography and the geography of man there is no longer a gap. Both are in the closest contact in terms of objects and methods. Thus in my opinion, the physiognomic approach is a great gain for geography, although thereby the field of its enquiry is greatly narrowed.

[1] Leo Waibel, *op. cit.*

Otto Schlüter

The significance of Schlüter's work to geographic scholarship is profound. All German geographers are agreed on this. It is remarkable that he made his methodological contribution sixty years ago, and that his greatest scholarly work occupied the whole of his life and was not published until shortly before his death. The measure of his importance is to be gauged not merely from his personal researches, but from the basic ideas he expressed in 1906. In other words, the influence of Schlüter's insistence on the study of the human habitat as the work of human groups is to be assessed from the trends of research throughout the first half of the twentieth century, whether it be in the study of individual landscape elements, such as field systems, farmsteads, urban types, or road nets; or whether in the spatial combinations of such visible elements of the landscape to form distinct compact associations to which the geographer has given the name of *Landschaft*.

Whether directly or indirectly, the ideas of Schlüter and of his contemporary Jean Brunhes were transferred to the research objectives of geographers. The geographic concept of landscape, however, made no impact in Britain between the wars, and is generally regarded with scepticism today. The book edited by H. C. Darby in 1936 on the historical geography of England[1] was a symposium, to which British geographers contributed fourteen scholarly essays. Darby includes a quotation from Brunhes in the preface, but in fact the essays make no contribution to a visualization of the landscape, either by the distribution of its elements, or by the determinants of their forms and spatial arrangement. The same was true in the United States down to the twenties, when Carl Sauer[2] sought to shift the focus of study to the landscape and the landscape unit and turned the attention of his colleagues to trends in Germany and the need for working out objectives of enquiry and methods of recording them in the field. When he left Michigan for Berkeley, he left behind a group who referred to themselves as chorographers and at that time were ardent students of Siegfried Passarge. One of their leaders was Preston E. James. They were in fact pursuing the tradition of Schlüter and Brunhes. Just as Hettner scorned and misinterpreted Schlüter's interpretation of landscape as 'aesthetic', so Wellington Jones (as I recall from contacts with him in Chicago, beginning in 1931) used to raise doubts about judging landscape in terms of its 'lookishness'. Such criticisms are set at rest by inter-

[1] H. C. Darby (ed), *An Historical Geography of England before A.D. 1800, Fourteen Studies*, Cambridge, 1936.
[2] Carl O. Sauer, 'The Morphology of Landscape', *University of California Publications in Geography*, 2 (1925), pp. 19–53.

preting the elements and spatial arrangement of landscape elements in terms of their functions, or, in other words, as it is often called today, morphogenesis. It is a pity that in America the study of landscape elements, such as farm types or urban structures, covered bridges, or unit areas of landscape associations, should have stopped after a brief period of enthusiasm in the early thirties. This is a tradition and purpose of geographic enquiry that needs to be revived and developed in both Britain and America. The tradition was digested and represented in Sauer's monograph on the morphology of landscape and one of the very few examples of such studies is Jan Broek's study of the sequence of human occupance in the Santa Clara valley. The study of landscape is a central theme in the studies of both French and German geographers, as to types, their distribution, their origins and spread, and their spatial association today as functioning entities in the organization of space. The long established close allegiance in both France and Germany of geographers and historians and cultural anthropologists is clear evidence of this. Among German geographers in the thirties, and their works in later years, we draw particular attention to the works of Oskar Schmieder and the late Leo Waibel and their followers. To their works we shall turn in a later chapter.

Hartshorne in the thirties concluded a long critical discussion of the limitation of geography to the 'visible phenomena' of the landscape, as advocated by Schlüter and opposed by Hettner in the following words:

The doctrine that the field of geographic study must be limited to areal phenomena which are 'in general, visible' admits of no precise statement other than a limitation to material features. Although enthusiasts have been preaching the doctrine for over thirty years it has been accepted by a relatively small minority of geographers; still fewer maintain it in practice. This is not surprising since it is founded neither in the logic nor in the historical development of geography, nor does it offer the basis for a more restricted though unified field. On the contrary, it would disrupt even a 'central core' of 'pure geography' by placing certain aspects of area outside of the geographer's field of study, except as he finds it necessary to investigate them in order to interpret what he has studied. While it cannot exclude those fields of geography which have ever been concerned with primarily immaterial features, it would thrust them into ill-defined outer reaches of geography with no better guide to their objectives than that of the study of relationships. By reducing the element of relative location to a purely secondary position it tends to neglect the very essence of geographic thought—integration of phenomena in spatial association. In consequence there develops an uncritical enthusiasm for patterns as such, regardless of their significance, and an overemphasis on form in contrast to function which tends to slight those characteristics of areas whose importance is not represented proportionately by material objects.[1]

It is far from my intention to enter a philosophical discussion of the relative merits of the views of Hettner and Schlüter as to the content and purpose of geography, the one focussing attention on terrestrial areas, the other on terrestrial landscapes. Their views, as expressed in published papers over fifty years ago, have been fully and critically discussed by R. Hartshorne in *Nature of Geography*, whose discussion is concluded in the paragraph quoted above. Suffice it to note here that for two or three decades Hettner's approach resulted in the orderly presentation of categories of data on the earth's surface viewed in their relation to a physical framework. Schlüter's approach focussed attention on the use of land by man in terms of his heritage and purpose and technology. Vigorous discussions and exhaustive research studies in Germany and France in the inter-war years (when Hartshorne's work was written) was directed to the question of reconciling these two modes of approach. Gradmann's exhaustive presentation on south Germany followed the Hettnerian mould, but the individual studies upon which it was based (as in plant geography and the lay-out of human settlements in Wurttemberg) sought to understand the areal variations of particular sets of phenomena in the way advocated by Schlüter and, incidentally, many scholars in other fields. The search for the distribution in the landscape of particular forms of land use—for example, idle patches of land or scraps of reafforested land—is a visible landscape phenomenon whose distribution was determined and then interpreted in the light of social and economic forces. This has been a major breakthrough of geographic research in Germany in the post-war years. Similar research trends are evident among historical geographers (such as Roger Dion) and medieval historians (such as Marc Bloch) in France. Indeed, the impact of the points of view of both these scholars is evident in the work of their successors. The debt due to both of them is recognized in the tributes paid to them by the present generation of professional geographers. Hartshorne's conclusions in the thirties must be assessed in the light of the work done since then, that is, over the past thirty years.

[1] R. Hartshorne's *The Nature of Geography*, 1939, p. 235.

Germany

Further References

There are two recent appraisals of the significance of Otto Schlüter's work. These are H. Lautensach, 'Otto Schlüter's Bedeutung für die methodische Entwicklung der Geographie', *Pet. Mitt.*, 96 (1952), pp. 219–31; and R. Käubler, 'Otto Schlüter's Bedeutung für die geographische Wissenschaft', *Die Erde*, 95 (1964), Heft I, pp. 5–15.

11

Contemporaries of the Second Generation

SIEGFRIED PASSARGE

Siegfried Passarge (1866–1958),[1] a contemporary of Hettner and a pupil of Richthofen, belongs to the second generation. His professional work extended over a period of nearly sixty-five years, for his publications began with a geological study of the New Red Sandstones in Thuringia in 1891 and ended with a review of morphological studies in the desert around Aswan in 1955. Passarge was a medical doctor and geologist and became a geographer. He was a prolific writer and his ideas are often so complex and contradictory as to be difficult even for a German clearly to interpret, yet he stands as one of the lone makers of modern geography. He received the Ritter medal of the *Gesellschaft für Erdkunde* in 1953 and an honorary doctorate from the University of Hamburg in 1956 shortly before his death at the age of ninety-two.

Born in Königsberg, where his father was *Amtsrichter*, Passarge entered the University of Berlin in 1886 where he was advised by Richthofen to train in the natural sciences as a preliminary to becoming a geologist. He soon moved to Jena and then to Freiburg. At the latter University he decided to take up medicine and spent four years at Jena (1888–92) in this training. At the same time, however, he pursued his geological study and in 1891 and 1892 he took the doctorate examinations in both fields. He spent his vacations at this time in excursions on foot, making geological observations, taking notes, and painting. In 1893 he was invited by a government body to the German Cameroons and made there observations of land and people that were published in 1895. On returning to Berlin, he attended Richthofen's *Kolloquium* and at the same time practised as a doctor in a hospital. After attending the International Geographical Congress in London in 1895, he returned in 1896 to

---

[1] On Siegfried Passarge, Helmut Kanter's article, 'Siegfried Passarges Gedanken zur Geigraphie', in *Die Erde*, 91 (1960), pp. 41–51.

Ngamiland in Africa to report to a British company on prospects of diamonds and gold. From here he moved to the Kalahari, where he became particularly interested in the geology, land-forms, and animal life. This work was incorporated in a book on the Kalahari, published in 1904, and on this work he gained his *habilitation* with Richthofen as his examiner. In 1901–2 he went on a sponsored visit to Venezuela and travelled far into the interior. In 1905, he was called to the chair of geography at Breslau that was vacated by Partsch who had moved to take Ratzel's place at Leipzig, and in 1906–7 he was in Algeria where he made landscape observations, published in 1909, that subsequently directed his approach to the geographical study of landscape. In 1908 he accepted a post in the newly established *Kolonial Institut* in Hamburg and here he remained for the rest of his professional life, retiring in 1936. During the First World War he served as a doctor and in 1917, while in Africa, had a serious illness.

I turn now to his professional work. He was strongly opposed to the Davisian approach to the study of land-forms that swept Germany after his lectures at the University of Berlin in 1908–9. This was expressed in *Physiologische Morphologie*, published in 1912,[1] where he exposed the weaknesses of the Davisian system, particularly from the standpoint of the geographer whose concern is the reality of the landscape, and where he first introduced the idea of *Landschaftskunde*, which he developed in extensive writings over the next twenty-five years. He grasped the validity and importance of Herbertson's conception of the 'major natural regions' (1905). Each major natural region, he pointed out, is a composite of natural phenomena, of which climate is the most important basic determinant, and each has its peculiar processes and forms. He also showed how these processes change through time, and have vestigial remnants in the existing land-forms. He sought to establish the hierarchy of physical units, the 'tissues and cells' as Herbertson called them, that make up the spatial content of the major natural regions. This approach and its procedure is clarified in *Die Grundlagen der Landschaftskunde*, three volumes, published in 1919–20 and *Vergleichende Landschaftskunde*, published in five parts, 1921–30. The major world regions, *Landschaftsgürtel*, he said, are essentially climatically determined, and within these there is a fourfold hierarchy of landscape entities. The field is clearly outlined in a brief introductory work published in 1923 (*Die Landschaftsgürtel der Erde: Natur und Kutur*, Breslau, 2nd ed. 1929).

---

[1] S. Passarge, *Physiologische Morphologie*, 1912, and an article with the same title in *Pet. Mitt.*, 58 (1912), pp. 5–8.

Contemporaries of the Second Generation

This was his springboard for examination of the meaning of cultural geography (*Kulturgeographie*) and the cultural landscape (*Kulturlandschaft*) in the thirties. He developed the idea of four spatial forces—*Raum, Mensch, Kultur, Geschichte*, and the idea that cultural geography is concerned with the influence of Man and his works in the creation of the cultural landscape or the man-made landscape. He urged that one of its themes should be the impact of the landscape on the psychology of the human occupants, and he coined and developed the idea of the *Stadtlandschaft*, writing a short book on this theme in 1930 (*Stadtlandschaften der Erde*).

He turned to two other themes in the later years of the thirties and forties, *Volkskunde* and *Länderkunde*—ethnography and 'regional geography'. He wrote a work on ethnography from a geographic standpoint in 1934 (*Geographische Völkerkunde*), and early in his career, he wrote three standard 'regional geographies', on South Africa (1908), on the Cameroons and on Togo (1914). He turned more assiduously to the conceptual basis of this field, rejecting the view of Richthofen's day that 'regional geography' was an orderly presentation of a large variety of spatial distributions. To this end he made a thorough study of the history of the idea among geographers from about 1700. This manuscript was lost during World War II, but we are told[1] that he discussed in it the ideas of *milieu* and *lebensraum* as embracing the totality of the spatially operating forces that contribute to the individuality of areas and along these lines he envisaged the distinctiveness of *Länderkunde*. These ideas are in line with the most recent developments in geographic thought and work and it is of interest to find here the convergence of French and German ideas and the clear distinction between *Landschaftskunde* and *Länderkunde*—the convergence of the ideas of Vidal de la Blache, Hettner, Schlüter, and Penck.

The study of landscape, Passarge argued as far back as 1912, in *Physiologische Morphologie*, involves the recognition and description of repetitive forms on the earth's surface. Referring first exclusively to land-forms, he claimed that it is impossible to define terrains (i.e. the areal variables of landscape) in terms of ONE genetic process as was assumed to be Davis' stand. There are, he claimed, both local and genetic forces, the first being geological, the second mainly climatic. Instead of the Davisian formula of 'structure, process and stage', Passarge wished to substitute 'Region, Place and Form', that is, the region, the local forces, and the resultant land-forms. He then proceeds to an analysis and classification of the form-elements (*Einzelformen*) grouped in terms of their genetic deter-

[1] H. Kanter, *op. cit.*

139

minants. If one genetic force is dominant, the resultant form of the land is called a 'monodynamic form', if two or more it is called a 'polydynamic form'.

Passarge then gives an example as to how, from the geographic standpoint, the land-forms of an area should be studied. A good topographic map of slopes (relief) is needed as a base. This must be accompanied by maps on the same scale of the lithology and vegetation. These, when superimposed, provide the physiological base for the recognition of the morphological types. He points out that this method of morphological study is fundamentally different from the 'doctrines' of Davis, and that the geographical study is embedded in the real face of the earth. The Davisian method gives an apparent explanation of land-forms, based on abstraction, whereas the explanation is often obscure and complicated and raises difficult problems of surface forms that the Davisian theory cannot answer. 'This method', argues Passarge, 'leads to quick and superficial conclusions that are highly dangerous to beginners.' So he concludes, 'If we do not succeed in checking the Davisian method, then the development of morphology could undoubtedly be permanently extracted from the field of geography and passed to geology. For the development of geographical science that would be a severe blow.'

Passarge's big work on the principles of description of landscape, published in the twenties, sought to develop a distinctive classification of the forms of the landscape and a procedure for explaining their characteristics.

His first major work, in four volumes, was called *Grundlagen der Landschaftskunde* (1919–20). Volume 1 considers the individual or unit elements of landscape, a listing of Teutonic thoroughness that obviously irked both Davis and Bowman in the States.[1] This however, was preliminary to the next volumes. Volume 2 deals with the causes of the landscape phenomena, sea, atmosphere, plants and animals; Volume 3 with the origin of land-forms (processes of erosion and deposition and their differential operation in different climatic regions); Volume 4 with Man's work as evidenced in the landscape.

A few years later in *Vergleichende Landschaftkunde* Passarge further developed the idea of the unit area and the hierarchy of such units. He drew a distinction between two kinds of landscape elements. The first are the individual groups of elements—climate, water, land, plants and cultural phenomena. The area of distribution is called an *Erdraum*. The landscape as an areal composition of these elements is described as a *Landschaftsraum*. The smallest areas are

[1] I. Bowman, 'The Analysis of Land Forms: Walther Penck on the Topographic Cycle', *Geog. Rev.*, 16 (1926), pp. 122–32.

the 'landscape elements', such as slopes, meadows or valley bottoms, ponds, dunes etc. A grouping of contiguous elements (*Landschaftsteile*) made up a section (*Teillandschaft*), and contiguous sections form a region (*Landschaft*). The latter again form contiguous groups called *Landschaftsgebeite* (e.g. the North German Plain); and a number of these make up a region such as the Central European Forest. This in turn is a constituent member of the world's great regional belts *Landschaftsgürtel* that traverse the earth in broad latitudinal belts, virtually identical with the climatic zones. The important point is that this system is a study of the visible landscape, including the fields and settlements and other imprints of Man, but the facts of the man-made landscape are *Not* considered as delimiting factors.

The compact geographical unit (*Teillandschaft*) is identical with Schlüter's idea of a landscape unit of perceptible phenomena, natural and human, that together form a single association. Schlüter, however, limited his study to the landscape—the origin, spread and association of its elements. Passarge went much further, for in subsequent works, he pursued fully, often obscurely, the precise nature and extent of the direct and indirect influences exerted by the *Landschaftsraum*, including its urban areas (*Stadtlandschaften*) upon *all* aspects of human and animal life. Thus, as he frequently stated, one is led into the realms of biology, psychology and medicine. It is precisely the limitless ramifications of this approach that Schlüter rejected.

Passarge's notion of landscape became a focus of interest in the twenties and thirties. This is evident in J. Sölch's remarkable essay published in 1924,[1] in which he defined any area, irrespective of extent, where there was a measure of homogeneity, a *chore*, from the old Greek word meaning area. A. Penck used the word but changed the meaning. He defines the smallest recognizable unit area as one which possesses unity of Form and Function. Such areas, he writes (in 1928) 'we shall call chores, adopting in a more limited sense the term used by Sölch'.[2] N. Krebs, the pupil and successor of Penck in Berlin, held the same view but made explicit that the social structure of society must be assessed together with the habitat which it creates and occupies. Leo Waibel sought an ecological base for the recognition of unity in the agricultural landscape.[3]

[1] J. Sölch, *Die Auffassung der naturlichen Grenzen in der wissenschaften,* Innsbruck, 1924.

[2] A. Penck, 'Neuere Geographie', *Sonderband Ges. f. Erdk. z. Berlin,* 1928, pp. 31–6.

[3] Leo Waibel, *Probleme der Landwirtschaftsgeographie,* Breslau, 1933, 94 pp. See also Ch. 12.

Germany

ALFRED PHILIPPSON

Alfred Philippson (1864–1955)[1] was one of the 'great masters of the classical period of our science', to use the words of Herbert Lehmann in an obituary appraisal. He trained under Richthofen at Leipzig and was initially, like his master, strongly founded in geology. His geographical work reveals the 'compartmentalization' or 'layering' of subject matter as conceived by Richthofen and Hettner, but the really remarkable feature of Philippson's work was the fact that it was concentrated in one area for over fifty years—the lands of the eastern Mediterranean. The period of his active research and writing covers 66 years and he was almost ninety years old when his first volume on the regions of Greece was published.

His work began with prolonged periods of travel in Greece between 1887 and 1904, when he collected both geological and geographical data over routes traversed mainly on horse-back with a total distance of about 20,000 kms. His first major *Landeskunde* was entitled *Peloponnes: Versuch einer Landeskunde auf geologischer Grundlage*, published in 1891–2. It contains 647 pages and four original maps on a scale of 1: 300,000 and the whole is based on personal observation and field mapping. This work of the 28 year old scholar, says Lehmann, carries all the traits of the master, Richthofen. It is 'ein Meisterwerk in seiner Art classisch', for Philippson had filled, through exploration, one of the large blanks on the geographic map, to which Richthofen had directed his pupil's attention. His geological and relief maps are still fundamental to the understanding of this area. He produced further volumes on Thessaly and Epirus in 1897, on the Aegean islands in 1901, and on western Aisa Minor in 1911–15. For all these studies Philippson prepared, single handed, 26 maps of the geology and relief on a scale 1: 300,000. In 1918 there was published his geological essay on Asia Minor (183 pages) in the German Handbook of Regional Geology (ed. Steinmann and Wilckens). This was a remarkable record, all of it achieved before the First World War.

He became professor at Bonn in 1911 and retired in 1929. A major work on general geography (*Grundzüge der Allgemeinen Geographie*), in three volumes was published in 1921, 1923 and 1924 (revised edition 1930, 1931 and 1933), and a major work on Europe, excluding Germany, appeared in 1928, successor to a first major work of 635 pages that had appeared in 1894. A book on the Mediter-

---

[1] H. Lehmann, 'Alfred Phillipson zum Gedächtnis anlässlich der 100. Wiederkehr seines Geburtstages am I Januar, 1964.' *Geog. Zeit.*, 52 (1964), Heft I, pp. 1–6.

ranean, 'its geographical and cultural personality' first appeared in 1904, and the fourth edition in 1922. In 1928 he resumed his field researches in Greece, and continued geomorphological enquiries. In 1934 he was honoured by the Academy of Sciences of Athens. He urged the need at that time for a definitive study of the geography of Greece, a work which he had begun to write in Bonn in 1929. In that year he retired. Thereafter, he wrote a major geographical work on a *cultural*, as distinct from a *geological*, basis. This is his little known, but important study of the Byzantine Empire in its geographic expression (*Das Byzantinische Reich als geographische Erscheinung*. Leiden, 1939, 214 pages). Phillippson also travelled on many occasions in Italy and published significant regional essays on Apulia (1937).

In 1942 Philippson, at the age of 78, with his wife and daughter, was deported to the concentration camp of Theresienstadt. He was there for three years and wrote an autobiography (*Wie ich Geograph wurde*) that is not yet published, but as Lehmann remarks, it throws much light on the development of geography in Germany. After the war Philippson returned to Bonn and it was there that I met him in his apartment in the company of Carl Troll in 1947.

His manuscript on the geography of Greece had been safely preserved in Bonn and the main effort of his life's work—*Die Griechischen Landschaften*—was continued, and completed by his two most devoted followers, H. Lehmann and E. Kirsten. Philippson's writing in his eighties continued, and, he wrote a masterful essay on Bonn (1947) and two essays in Carl Troll's new periodical *Erdkunde* on Greece (1947) as well as a long work on the climates of Greece (1948).

The crowning work is the *Griechische Landschaften*, that was published in the fifties. This monumental work covers four volumes and nearly 3,000 pages of text and was written some 25 to 30 years ago. The first volume deals with Greece as a whole, the second with the Northwest, the third with the Pelopponese, and the fourth with the Aegean Sea and its islands. It ranks as the standard work on the geology and geomorphology of that country, but is obviously well behind the times, especially in its human aspects, in view of the researches (particularly in French) that have been conducted in that part of the world in the last twenty-five years. Yet it stands as an exemplary scholarly exposition of the concept and procedure of regional geography (*Länderkunde*) in the light of the ideas and teaching of Richthofen. This work achieved essentially what the young scholar had sought for in his work on the Pelopponesus sixty years earlier: a *Landeskunde* on a geological-morphological base. Its objective is as formulated by Philippson sixty years earlier: 'to characterize the naturally bound earth-area as a whole, and all the individual objects that contribute to its individuality'. In this work on the Pelopponesus, in which this statement appears, its author recognises that many of the phenomena in the landscape are rooted in culture and bear no causal relation to the 'geological-morphological' base. It is significant that the German archaeological institute took an active interest in Philippson's work throughout his long professional life, and he turned increasingly towards the study of the man-made existing landscape in his later studies.

ROBERT GRADMANN

Robert Gradmann (1865–1950)[1] came to geography through botany, and the faculty of Natural Sciences of the University of Tübingen awarded him a doctorate degree for work on the plant life of the Swabian Alp. He thus combined in an unusual way a training in natural science and history, and bridged the gap between study of the natural landscape and the beginnings of cultural history, through which there emerged the cultural landscape. In so doing, assert his colleagues on the occasion of his 85th birthday, he became 'a real geographer'. In the unusual association of plant geography, pre-history, and land settlement he was able not only to reveal the development and changes of the central European landscape, but also to create new foundations for the further investigation of the geography of settlement. In his researches on rural and urban settlements in Württemberg he blazed new trails. His researches gave new points of view to the geographic aspects of cultural and economic history and particularly to agrarian history.

Gradmann, depending primarily on botanical evidence, developed a theory of the distribution of a so-called steppe-heath formation at the beginning of the period of effective sedentary occupance in the Neolithic period. His starting point for this theory was based on the botanical vestiges of this plant association that are to be found in south Germany. He postulated that at this period there was a strong contrast between the forest-free areas with a steppe-heath and open woodland vegetation, and the dominantly forested areas. The forest-free areas were the seats of the earliest finds of Neolithic settlement. He first assumed that Neolithic groups settled these steppe-heath

---

[1] A tribute to Robert Gradmann by his colleagues on the occasion of his 85th birthday. 'Robert Gradmann zum 85 Geburtstag', *Berichte für Deutschen Landeskunde*, 8 (1950), Heft 2.

areas during a dry warm period. He later (1930) modified this to include the areas with calcareous soils, which were better suited to a system of economy involving arable fields and rough pastures without manuring, with wheat and barley as the main crops.

His greatest achievement was the two volume work on south Germany. This is regarded as a standard work of *Länderkunde*. He had previously brought together numerous studies for this composite work in the description of a number of administrative districts (*Oberamtsbeschreibungen*) and in the total presentation for the kingdom of Wurttemberg. The tradition, it seems to us, cannot be separated from the current series of such studies that is being sponsored by the *Bundesamt fur Landeskunde*. In writing these 'descriptions' he raised and tackled a variety of questions, such as the origins of limestone land-forms, the problem of scarp formation, and the glacial formation of Lake Constance. His two-volume work *Süd-Deutschland*, published in 1931, is the most reliable guide in print for its wealth of fact and the diversity of its references from many fields of enquiry. This, however, is not only the work of an assiduous bibliophile. Gradmann was a great field worker and an artist. He was opposed to all formalism, but he frequently indulged in theoretical discussion of the objectives and methods of *Länderkunde*, and he introduced the concept of 'the harmonic landscape' into geography, which raised much discussion between the wars. We also recall his discussion of the relations between plant geography and the history of human settlement—his papers on plants and animals in the conceptual field of geography in 1906 and 1919. His concept of *Steppenheide* raised many enquiries and much controversy in the thirties and he pursued the same theme in personal observations in Algeria, the edge of the Sahara, and the Orient.

Gradmann took his *habilitation* on the special wish of Karl Sapper at the University of Tübingen, where he was employed in the University library, as well as being employed in the statistical office of Württemberg. In 1919 he accepted a call to the University of Erlangen and it was there that he wrote his classic work on southern Germany and also established the institute for *Fränkische Landesforschung*. He edited the German research series in geography, (*Forschungen zur deutschen Landeskunde*), and made contributions to it on the rural and urban settlements of Württemberg. The University of Tübingen awarded him the honorary degree of doctorate on his 76th birthday. The Geographical Society of Munich awarded him their gold medal and the Geographical Society of Berlin awarded the Carl Ritter medal.

Germany

KARL SAPPER

Karl Sapper (1866–1945)[1] began his studies in geology in Munich and his geological competence is reflected in his subsequent work. But he was early drawn to geography by two circumstances; familiarity with his home area in the Swabian Alp, and his departure for Guatemala on account of poor health immediately after the completion of his academic studies. In 1888, at the age of 22, he joined his brother, a coffee planter, in Guatemala. He ran a plantation in the rain forest for two years, but again had to give it up because of ill-health. Here he learned by direct experience the rudiments of tropical farming, the manners of the local Indian people (he learned their speech, since they did not use Spanish), and carried out topographic and geological surveys of the area in which the plantation was located. He also took an interest in native ethnography and archeological finds. After two years, he moved to the town of Coban, and, thanks to his brother's support, he was able to dedicate himself to his scientific interests. He travelled extensively throughout central America, and for a time served the Mexican government as a geologist. Most of his travel was done on foot, accompanied by two or three Indians. He kept regular diaries and made consistent observations on geology, morphology, weather, vegetation, and economic and social conditions. He published his work as he collected and soon became an acknowledged authority on central America. He prepared geological maps of central America and a large topographic map of Guatemala.

In 1900 Sapper returned to Germany at the age of 34, after twelve years of scientific field work, a record that is rare indeed among scholars of any period. He was encouraged to take his *habilitation* by both Ratzel and Richthofen to enable him to lecture in German Universities. In 1902 he became a professor of geography (*extraordinarius*) in Tübingen. Here he stayed until 1910. In that year he accepted a call to Strasbourg, where he succeeded G. Gerland. In 1919 he moved again to Würzburg, where he built up a new institute of geography and remained there until his retirement in 1932.

While in central America Sapper became very interested in vulcanology and F. Ratzel encouraged him to write a book on this general theme. Sapper needed more first-hand observations to fill this task to which he committed himself, and the observation of volcanoes was the primary purpose of his travels in the next ten years until 1914, particularly in the Indies and the Mediterranean lands. He undertook two great journeys in the twenties—one in

[1] F. Termer, 'Karl Sapper', *Pet. Mitt.*, 91 (1945), pp. 193–5.

146

Henri Baulig

Raoul Blanchard

Jean Brunhes

Albert Demangeon

Plate 1

Vidal de la Blache

Plate 2

Emmanuel de Martonne

Robert Gradmann

Hugo Hassinger

Alfred Hettner

Plate 3

Alexander von Humboldt

Hermann Lautensach

Joseph Partsch

Albrecht Penck

Plate 4

Alfred Philippson

Plate 5

Friedrich Ratzel

Plate 6

Ferdinand von Richthofen   Carl Ritter

Plate 7

Otto Schluter

Maximilien Sorre

Carl Troll

Leo Waibel

Plate 8

central America, and the other throughout south America as far south as Buenos Aires.

The published works of Karl Sapper are enormous and only the general features can be noticed here. His earlier works are on physical geography and especially on geology, and on the geography of Central America, which he knew probably better at first hand than anyone else. He also made outstanding contributions to the cartography of central America. His work on vulcanology occupied his life-time and did not appear until 1927. He took a strong interest in the tropics in general and in problems of economic development and acclimatization. These interests appear in a book on the tropics—*Die Tropen*—that was published in 1923. His ideas in economic geography are systematically presented in a work on general economic and transport geography (*Allgemeine Wirstschafts—und Verkehrsgeographie*) that appeared in 1930 shortly before his retirement.

Karl Sapper was one of the most widely travelled workers of the second generation of German geographers. An appraisal in 1945 states: 'As a researcher and a man German geography will long count him as one of its best representatives.'

One inevitably must pause to make several closing comments on the career of this remarkable scholar. Unlike any geographer in the English speaking world, he was able to spend many years of first hand observation, with the same objectives and methods and in the same general area as his illustrious predecessor, Alexander von Humboldt. There is a strong and long tradition in German geography of extensive, arduous and dangerous travel in the wildlands of Latin America, that has involved the accumulation of daily records, mapping and publication of the results. This is evident in the works of Hettner, Schmieder, Waibel, Wilhelmy, Troll, Pfeifer and others. These men have a distinguished predecessor in Sapper, a senior colleague of the second generation, who, like other German geographers, spent many years in the field, away from their academic studies, before entering University work in his native land.

HUGO HASSINGER

Hugo Hassinger (1877–1945)[1] died from a road accident in his seventy-fifth year. His work is little known or appreciated in the English speaking world, but it must receive special emphasis here. Hassinger was one of the great geographers in the first half of this

---

[1] E. Lendl, 'Hugo Hassinger und die landeskundliche Forschungsarbeit', *Berichte für Deutsche Landeskunde*, 13 (1954), Heft 1, pp. 1–10.

century. He held the chair in Vienna and all his work was in central Europe and is all published in German. He studied under Penck in Vienna, together with contemporaries such as Norbert Krebs and Robert Sieger. He was born in Vienna and studied geography, geology and history at the University. His first research (1905) was on the geomorphology of the basin of Vienna, but he soon became more interested in the city itself. He took his *habilitation* in 1915 on the cultural geography of the Moravian Gate area and on the strength of this became a *privat dozent* in the University of Vienna. Throughout his life he was dedicated to the problems of human geography in central Europe. He moved to Basel after the First World War and remained there until 1928. He then followed Krebs at Freiburg and remained there for three years. He then returned to Vienna to hold the chair of anthropogeography, in succession to Tomaschek and Oberhummer.

He directed regional studies while at this post, and he made early studies of Vienna (1910), its location and morphological structure, and published the results as an atlas in 1916. He also continued work on the eastern section of Austria from the standpoint of human occupance. He paid attention to the production of regional atlases and was responsible for the atlas of Burgenland and he published in 1917 an excellent geographical portrayal of *Mitteleuropa*. His other major works include the section on anthropogeography in Klute's *Handbuch* in 1938 and on the geographical foundations of history in 1931; but his major individual research contributions lie in his regional studies in Austria and southeastern Europe.

ALFRED RÜHL

Alfred Rühl (1882–1935)[1] was professor of geography for many years in the University of Berlin, where he was attached to the *Institut für Meereskunde* in the Institute of Geography. He was thus a colleague of Albrecht Penck, who was actually responsible for recommending Rühl's appointment. Rühl was a rigorous and productive worker in the field of economic and social geography, and was, as it now proves, ahead of his time. Neither his name, and certainly not his works, are even familiar to English speaking geographers,

[1] For Alfred Rühl see the following—O. Quelle, 'Alfred Rühl', *Pet. Mitt.*, 81, (1935), pp. 368–9; K. Oestreich, 'Alfred Rühl', *Geog. Zeit.*, 42 (1936), pp. 143–7. R. Steinmetz writes an introductory appraisal to Rühl's book on *Einführung in die allgemeine Wirtschaftsgeographie*, 1938. This book also contains a bibliography of Rühl's writings (pp. 85–93).

and yet he represents an important trend in the development of modern geography, that has become a dominant theme in geography since his death. It is for this reason that I shall briefly discuss the place of Alfred Rühl in the growth of geography in Germany.

Born in Konigsberg in 1882, Rühl studied geography and the natural sciences in Königsberg, Leipzig and Berlin. He furthered his formal studies in Berlin and Marburg where he was appointed in 1909 as assistant to Th. Fischer. He accepted a post at the Institute of Oceanography in Berlin, under Penck, in 1912, and in 1914 he became the second professor in Berlin alongside Penck. Shortly after the First World War, he received the status of honorary professor in the *Tecnische Hochschule* in Berlin and attained the rank of professor in Berlin University in 1930. He suffered from ill-health and died in Switzerland in August, 1935, at the age of 53.

Rühl was Richthofen's last doctoral candidate and his thesis, submitted in 1906, on the morphological effects of ocean currents on the Mediterranean coasts, was inspired by Richthofen. He continued his work on the Mediterranean lands, mainly on geomorphological themes, and this was a reason for his going to Marburg, for Fischer at that time was the recognized authority on that area. The culmination of Rühl's geomorphological work was his translation from English of Davis' lectures on the 'explanatory description of land-forms', published in 1912. This book had a strong influence in Germany and Rühl was always regarded (with Gustav Braun) as an outstanding supporter of the Davisian approach.

He now turned to the field of economic geography with which his name is particularly associated. In 1912 he took part in the transcontinental excursion in the United States under the leadership of his friend, W. M. Davis. During this visit he carried out investigations, that were subsequently published, of San Francisco and Newport News. It was Penck's purpose that Rühl should turn towards economic aspects of man's association with the oceans, and this probably accounts for his shift of interest. In 1920 his monograph on the foreign trade of German ports was published by the *Institut für Meereskunde*, and in 1926 another study was printed in the same series on the role of location in agricultural geography illustrated from eastern Australia. A special series of monographs were published in the late twenties dealing with the *Wirtschaftsgeist* of the Spanish, Orientals, and Americans. He intended to continue the series with studies of the Italians, Dutch, English and Chinese. Thus, eventually, he planned to arrive at a 'typology of economic man'. In these three published monographs he revealed an incredible width of reading and a distinctive approach.

Rühl's main recurring argument was that a geography of human economies must be based on a firm knowledge of the social sciences, especially economics, whereas in his day (and still often to this day) the training of geographers was confined almost entirely, in the Richthofen-Penck tradition, to the natural sciences and especially geology, with the result that the locale of economies was frequently attributed to natural rather than to economic forces. His approach is made most clear in a posthumous work published in 1938 with the title *Einführung in die Wirtschaftsgeographie*, with a long preface by R. Steinmetz, the distinguished human geographer in Amsterdam. As a harsh critic of 'geographical determinism', Rühl insisted that in the study of economic geography, deductive reasoning is necessary but inadequate. Hypotheses must be formed, but they cannot be rigorously applied owing to the complications due to a great variety of spatial variables. The highest level of certainty in geography is to be reached through induction, that is, through widening horizons of generalization from specific local areas—a principle that was clearly enunciated and followed by Ritter, and was the purpose of Rühl's 'geopsychological monographs'.

Rühl's last work needs special note. It was on the 'international division of labour: a statistical study on the basis of the imports of the United States'. This was published in 1932 and reveals the thoroughness with which Rühl was able to set a problem and use statistical data to answer it. He sought answers to such questions as— What are the factors that condition a country's imports? What quantitative relations are there between the various imported items? What are the connections between imports and exports on the one hand and domestic production on the other hand? He used American trade data for 1927 and came to the conclusion that only a little more than ten per cent of the U.S. imports came into direct competition with its own domestic production.

Rühl's oft-repeated plea was that 'geography must be a distinctive science like physics or botany'. To this end its field must be limited and its methods more rigorous. Rühl sought to do this in the realm of economic geography as in physical geography. Incidentally, the same judgement has recently been passed by H. Lautensach in his major work on the 'regional' divisions of the Iberian peninsula, in an entirely different geographical context.

It is of interest to note that in the late thirties, the most outstanding and appreciative of the appraisals of Rühl's work were by two Dutch geographers, R. Steinmetz, a human geographer from Amsterdam and K. Oestreich, a geomorphologist from Utrecht. Since then his work seems to have receded—it has nowhere even received mention in English literature. However, in 1960 a geographer in the East German Republic sought, rightly or wrongly, to argue that Rühl's approach was dominated by a socio-economic viewpoint closer to the communistic political philosophy than to western capitalistic philosophy, wherein, it is claimed, and I believe quite wrongly, that the spatial variants of human economies are still primarily attributed to the dictates of Nature, rather than to the aims of social philosophy. This is a grossly distorted interpretation of recent research trends in West Germany, and, as in so many other matters, seems to be a generation behind western views. That Rühl was strongly anti-Nazi can be said without doubt, but from that one cannot draw the conclusion that he was a communist. E. Otremba, the leading younger economic geographer in Germany, who is directly attacked by this communist colleague, remarks in a personal letter that Alfred Rühl is highly esteemed in the advancement of geographic thought in West Germany, particularly in the realm of human economies. Thirty years ago, Oestreich concluded: 'So ist er ein Anreger gewesen, dessen Arbeit weiter leben wird . . . Vielleicht war es hierzu noch zu früh'. This assessment is confirmed in a personal letter from Carl Troll who was for a short time in the early thirties a colleague of Rühl in Berlin. These assessments are now generally accepted among geographers, and we would instance the views of W. Hartke and H. Bobek, both of whom came under Rühl's influence while in Berlin, and there is much evidence of the same development in post-war trends in other countries. Even though workers may know nothing of the work of Rühl, his ideas are apparent in parallel developments of quantitative analysis in particular in both Germany and France, as well as in Britain and the United States. The danger is that these current trends may lean too far in the direction of the social sciences, as was the case fifty years ago towards the physical sciences, and towards geology in particular.

Further References

The following works have appeared since the book went to press: Franz Termer, *Karl Theodor Sapper, 1866–1945; Leben und Wirken eines deutschen Geographen und Geologen*, Lebensdarstellungen Deutscher, Nr. 12, Naturforscher, Leipzig, 1966, 89pp. K. H. Schroder, *Robert Gradmann: Lebenserinnerungen: zur Wiederkehr seines Geburtstages*, Stuttgart, 1965, 164pp.

12

Leaders of the Third Generation

The leaders of the second generation of German geographers retired in the late twenties. This was true of Penck, Hettner, Schlüter and others. This also happened to be about the time of the rise of the Nazis. The question then arises, who were the leading figures of the third generation from about 1933 to 1945? Several of the leading men were killed during the war. Among these were Hans Dörries (Münster), a pupil of Meinardus in Göttingen and Hans Schrepfer (Würzburg). But three scholars exercised a powerful influence in this period—Oskar Schmieder, Leo Waibel, and Wilhelm Credner. N. Krebs would also seem to fit in here, since he was an early pupil of Penck in Vienna, and ultimately his successor in Berlin.

NORBERT KREBS

Norbert Krebs (1876–1947),[1] an Austrian, went to school in Vienna and studied at its University under three of the greatest European scholars of their generation—Penck (geography), Suess (geology) and Hahn (climatology). He wrote his first dissertation on part of the Austrian Alps in 1903 and prepared his *habilitation* on the peninsula of Istria in 1907. Krebs first taught in a school for some years in Trieste, but in 1909 he became *privat dozent* at the University of Vienna. He worked alongside Ed. Bruckner and Eugen Oberhummer after Penck left for Berlin. In 1913 he published a major work on the regional geography of the Austrian Alps, and this was later extended and published again with the title *Die Ostalpen und das heutige Österreich* (1928).

Krebs was soon called to Germany and spent the rest of his life there. In 1917 he accepted a call to Würzburg and then to Frankfurt and then to Freiburg, where he stayed for seven years. He succeeded

[1] An obituary notice to Norbert Krebs in *Erdkunde*; H. Hassinger, 'Norbert Krebs zum Gedächtnis', *Erdkunde*, 2 (1948), pp. 200–2.

Penck at the University of Berlin in 1926. His long stay in southwest Germany is reflected in a three volume work on that area. While in Berlin he carried out studies relevant to the unity of the German lands (in which one suspects not only his Austrian beginnings but also the influence of his master, A. Penck). His book on *Deutschland und Deutschlands Grenzen* appeared in 1929 and he devoted the thirties to the supervision of the magnificent atlas of the *Deutsches Lebensraum in Mitteleuropa*, a project that was again initiated by A. Penck. In 1931–2 Krebs travelled to India and out of this emerged what is often considered to be his best work *Vorderindien und Ceylon*, published on the eve of the outbreak of war. Krebs retired in 1943, was bombed out of his home, retired temporarily to a village near Vienna, returned to Berlin in 1946, and died shortly after.

LEO WAIBEL

Leo Waibel (1888–1951)[1] has been named by a German colleague, Gottfried Pfeifer, as one of the greatest geographers of the inter-war period. Waibel trained under A. Hettner at Heidelberg and also spent a short time with F. Jäger, the specialist on Africa at Basel. He became an assistant to Thorbeke at Cologne and first visited the German Cameroons with his chief (1911–12). He was in German Southwest Africa at the outbreak of World War I and was interned there for the duration. He returned in 1919 to become *privat dozent* for one year at Cologne, and then moved to Berlin in the same capacity for one year, and became professor at Kiel in 1922. He spent a long research period in the field in Mexico in 1925–6 out of which emerged his study of the Sierra Madre de Chiapas, in which his ideas on economic geography were first developed. He remained at Kiel for seven years and then was called to the professorship at Bonn in 1929 where he remained until 1937. He then left Germany for the United States and remained there during the war years, moved to Brazil for five years (1946–50), spent another year in the States (1950–1), and returned to Bonn in 1951. He died on the eve of his return.

Waibel was greatly influenced not only by Hettner in his formative years but also by Albrecht Penck and Otto Schlüter. His distinctive stamp was left most markedly in the realm of economic geography, and especially the geography of agriculture, which he directed to the analysis of spatial systems rather than to crop and other individual

[1] For Leo Waibel see a short, but important appraisal by one of his oldest pupils G. Pfeifer, on the occasion of Waibel's 75th birthday, in *Geog. Zeit.*, 51 (1963), Heft 4, pp. 265–7.

distributions. The works of J. H. von Thünen, Eduard Hahn and T. H. Engelbrecht became known to him during his work at Heidelberg under Hettner. He early sought to reach a compromise between the conflicting views of Hettner and Schlüter, which he achieved through the pursuit of Albrecht Penck's famous dictum *Beobachtung die Grundlage der Geographie*—the theme of Penck's inaugural lecture on entering the University of Berlin in 1906. Waibel regarded Penck with the highest esteem—*Fürst im Reiche des Geistes*. While at Cologne, Waibel also came under the spell of the famous economic historian, Bruno Kuske, who was interested in the study of the changing spatial structure of economic spatial systems, as for example the role of Cologne in the geo-economic organization of northwestern Germany. From these varied sources as stimuli, Waibel endeavoured to advance geography—his own work as well as that of his students—by seeking for the relevant repetitive features of the visible landscape that reflect the spatial variants of human activities and attitudes. This was a combination of the ideas of Hettner, Schlüter and Penck. He developed the notions of *Wirtschaftsformation* and *Lebensform*, and on this basis formed what G. Pfeifer calls a *Waibelsches System*.

'Every inhabited area has its peculiar type of economy. To this corresponds a particular economic landscape, which is determined by the character of the operating unit of the economy (*Betriebsform*), and the aim of its production'. Thus according to Waibel, a particular kind of farming has a distribution in area which constitutes a farming and agricultural entity. It is part of a wider range of farming practices, which together consistute a major type of rural economy (*Wirtschaftsform*). Each of these types of economy is associated with a corresponding and distinctive economic landscape, itself an association of cultivated plants introduced and controlled by man, and called by Leo Waibel 'an economic formation'.[1]

There is the closest analogy here with the ecological objectives of the plant ecologist. These objectives and the methods are reflected in the works of his students in Schleswig and northwest Germany. The most distinguished today are J. Schmithüsen, a foremost authority on plant geography, with an interest in theoretical questions of the conceptual structure of geography; and W. Müller-Wille, a cultural geographer and specialist on northwest Germany. Waibel's earliest student (at Kiel) was G. Pfeifer, who later became his colleague at Bonn and today, in Heidelberg, is one of the leading geographers in Germany. H. Wilhelmy (Tübingen) also came under Waibel's (and Schmieder's) spell at Kiel, a fact that is clearly revealed

---

[1] L. Waibel, *Probleme der Landwirtschaftsgeographie*, Breslau, 1933.

in his early studies of rural settlements in Bulgaria and of the city of Sophia. Wilhelmy, now at Stuttgart, is now mainly concerned with geomorphology, but his interest in the geography of urbanism is evident in a book on the cities of Latin America, the major area of his travel and research.

The Tropics was Waibel's field of investigation, in the field and in the library. He first worked in Africa, beginning in the Cameroons with F. Thorbeke before the first war (1911–1912) and produced a book, after his return from internship in South Africa, on the natural resources of Tropical Africa (1937). After leaving Germany in 1937 he transferred his activities to Brazil, where he had a great impact on geographical colleagues, and produced important studies from field work, especially in the southern provinces in areas of German settlement. He always envisaged the production of a general work on the geography of the Tropics, but his specific professional activities and field work kept him fully occupied until his death in 1951. Waibel's ideas and influence thus reached three continents.

WILHELM CREDNER

Wilhelm Credner (1892–1948)[1] was one of the most productive and able of the third generation of German geographers, but he died at a relatively early age and his name, let alone his works, are little known in the English speaking world.

After World War I he studied geography under Hettner, following short spells at Greifswald and Upsala in Sweden. Like his fellow students, Schmitthenner and Jäger, he worked on a geomorphological problem for his doctorate—the land-forms of the crystalline rock areas of the Odenwald and Spessart. Geology was acquired through family tradition and he began life by training as a mining engineer. In his doctorate work he delved with rare insight into questions of the origins of peneplain and scarp in southwest Germany. For four years, from 1927 to 1931, he worked in Siam and south China, and here he continued his geomorphological bent by estimating the relative importance of the morphogenetic factors in the land-forms (published in 1931 in the Bulletin of the Geological Society of China and in a book on Siam in 1935). These ideas laid the foundations of modern climatic morphology. Yet Credner forsook these early contributions to geomorphology for the field of economic geography.

[1] C. Rathjens, 'Wilhelm Credner: Gedanken zu seinem 70 Geburtstag', *Geog. Zeit.*, 51 (1963), Heft 2, pp. 81–9.

His interest in economic geography was undoubtedly due to the influence of the thinking of Leo Waibel, another of Hettner's students. Credner left Heidelberg to join Waibel as an assistant at Kiel, and he took his *habilitation* with a work on the landscape and industry of Sweden (*Landschaft und Wirtschaft in Schweden*, 1925), becoming *privat dozent* at the age of 30. The ideas of Waibel on 'economic formation' and 'economic region' were developed by Credner. It is interesting to note this common interest, although it was derived from quite different backgrounds, Waibel from biology (notably zoology) and Credner from geology (mining). Credner wrote many works on mining in its geographic aspects—the ores of Sweden, the gold producing areas of the world, iron ore deposits of the United States, but death stopped his intention to write a work on the general geography of mining—deposits, production, and movements. In this interpretation, he was seeking to evaluate not only the physical factors of production, but also the relevant social conditions, such as access to labour supply, transport facilities, etc.

He never wrote a theoretical discourse but his ideas are evident in substantive works. Rathjens writes: 'Like Waibel, he placed the economic region (*Wirtschaftslandschaft*) in the centre of his research. This he regarded physiognomically in its regional expression (*landschaftliche Ausdruck*) and ecologically, according to both natural basis and economic forces. Economic regions were for him areas (*Raüme*) with the same economic imprint (*Prägung*). He sought to delineate these not only as to boundaries, but also as to typology.' His last works in this connection were on the West Indies and the agricultural regions of the United States. This approach, derived from the biological analogy of Waibel, he furthered by emphasizing the need for evaluating the historical development of such a spatial association in addition to the physiognomic and ecological aspects. It is not surprising that Credner laid emphasis on the study of economics for the geographer and that his chair was in economic geography at the *Technische Hochschule* in Munich. It is a short cut from this developmental approach (*Diagnose*) to spatial planning (*Prognose*) with which he was much concerned.

Credner was an expert at first hand on wide areas of southeast Asia and his book on Siam is a classic of regional literature. The same holds for his early work on Sweden, which was completed before his departure for the East. He also demonstrated the value of the geographer's approach to the problems of the developing areas, which, until a generation ago, were customarily described as 'colonial lands'.

Agricultural geography felt the benefit of his productive academic enterprise. This undoubtedly began through his contact with Waibel in Kiel. When placed in Munich in 1932 Credner continued this work, and began the first land use maps on a scale of 1:5,000 (with the help of Schmithüsen). This began through the *Landwirtschaftsgeographische Arbeitsgemeinschaft* established in 1937. He also aspired to an atlas of German field systems, a scheme that has been taken up since the war by a leading economic geographer today, E. Otremba, as the atlas of German agricultural regions (*Atlas der deutschen Agrarlandschaft*).[1] The objectives of Credner's large scale land use maps were explained in two articles in 1938 and 1942. He aimed not merely at the recording of land use, but also at the spatial variants of the whole agricultural structure (*Agrarstrukter*)—open- or hedge-bound field systems (*Zelgensystem*), parcel distribution, size of holdings, rotations, ownership, and distance of parcels from farmstead. Such work was also undertaken by Waibel in his institute in Bonn, many sheets being on the scale of 1:25,000. This has been continued since the war by W. Hartke in Munich and by Troll in Bonn, building on the pattern set by their predecessors.

In conclusion, Credner ranks among the masters of regional geography or *Länderkunde*. His work on Siam (1935) is dedicated to Hettner and still ranks as the standard work on this country. Between receiving the doctorate in Heidelberg and his death in October of 1948 there elapsed only 26 years, but Credner was a strong link in the chain of geographic scholarship in Germany, through Waibel back to Hettner, Schlüter and Penck.

OSKAR SCHMIEDER

Oskar Schmieder (1891–   )[2] was honoured on the occasion of his seventieth birthday in 1961 by a volume of essays on the geography of the New World. The contributors include 'friends, pupils and

[1] The first part of the atlas of the West German agricultural landscape was published in 1962. The work is being edited by E. Otremba, but it is being prepared by geographers and others in many Universities. Supported financially by the *Deutsche Forschungsgemeinschaft*, it is one of the greatest professional enterprises in Germany in the post-war years. The whole atlas is to fall into five parts—general maps of land use systems; general maps showing the relations between land use and physical condition; land uses on the urban fringe, illustrated by the case of Hamburg; the relevant aspects of social structure in their historical development and present forms; and representative large scale maps of field systems in different markedly contrasted areas of Germany. For a general review see the comments by Th. Kraus, on the first *Lieferung* of the Atlas, *Erdkunde*, 19 (Nov., 1965) p. 337.

[2] For Oskar Schmieder, see W. Lauer's introduction to the *Festschrift*, Vol. XX, *Geog. Studien*, Univ. Kiel, 1961.

Germany

colleagues', among whom are W. Lauer, C. Schott, H. Wilhelmy, H. Schlenger and H. Blume, who are among the leading geographers of the present generation. I chose to visit the University of Kiel's institute, under Schmieder's leadership, as one of the most active and distinctive in Germany in 1936. The following data are drawn mainly from Professor Lauer's opening appraisal in the Schmieder *Festschrift*.

Schmieder studied at Königsberg, Bonn and Heidelberg. His doctorate was completed under Hettner at Heidelberg in 1914, the topic being a geomorphological study of the Sierra de Gredos in Spain. Immediately thereafter he departed for Chile, his first visit to the New World, but he was obliged to return to Germany quickly owing to the outbreak of hostilities. He served right through the war and then resumed his academic studies at Bonn where he took his *habilitation* under Philippson. Almost at once he decided to go back to South America. He accepted an invitation to the University of Cordoba in Argentina as professor of mineralogy, and while there undertook several extended journeys. He was invited in 1925 to the University of California at Berkeley by Carl Sauer, a new arrival himself on that campus. After several years in Berkeley, Schmieder in 1932 accepted the chair of geography at Kiel. Shortly thereafter three volumes were published on the geography of South (1932), North (1933), and Central (1934) America; these were apparently written during his sojourn in California. He travelled subsequently in Argentina and the Gran Chaco, together with his younger colleague, H. Wilhelmy. In 1940 he served as chairman of the *Geographentag* and for five years (throughout the war) he was chairman of the new organization of the *Deutsche Geographische Gesellschaft*, which actually he called into being, and out of which emerged three volumes dealing with geographic aspects of current problems (*Lebensraumfragen europäischer Völker*, 1940, 1941, 1943). He moved briefly to Halle in 1944. He lost all his material possessions at the end of the war, his house and library in Kiel, and settled permanently in a summer house outside the city. In 1951 he represented West Germany at the celebration of the 400th anniversary of the University of Mexico. He then began to rewrite his work on the New World, for which reason, after he retired in 1956, he accepted a Visiting Professorship in Santiago, Chile (1958–9), so that he could bring up-to-date his knowledge of a land that had much changed since his first visit in 1914.

Schmieder has also travelled widely in the Old World, in north Africa and in Pakistan. In the latter country he was called to establish and launch a new department of geography in the University of Karachi. All told, he has spent some 15 years of his life outside Germany, 13 of them in the New World. He has learned thoroughly the language, culture, and attitudes of other peoples, especially in the Americas. He has a remarkable command of English and Spanish and has many student-followers throughout south and central America.

Schmieder began his studies, like most of his contemporaries, with geomorphology and the other natural sciences. But he early shifted to a more effective evaluation of the human element. One associates Schmieder particularly with the idea of the focal theme of geography as the transformation of the original natural landscape to the humanized habitat or the cultural landscape, a conceptual procedure that obviously calls for historical interpretation. W. Lauer writes as follows—'The development (*Inwertsetzung*) of the earth, transformation of the natural to the cultural landscape, the development history of geographical areas to their present aspect, are all objects of his works and of those who have worked with him.' This approach characterizes his work on the New World, in which he deals historically with the landscapes within the general framework of major geographical areas, but without serious concern for the definition of their boundaries or of their divisions. 'With this method he advances beyond hitherto accepted ideas, also in part beyond those of his master, Hettner. One should rather regard his three volumes on the New World as an attempt to apply the historical procedure of Robert Gradmann in southern Germany (whose views Gradmann had already outlined as far back as 1901) to the whole of a continent.' Thus writes Lauer. This is confirmed in Schmieder's own words in *Die Alte Welt*, 1965, p. xv. He seeks for a 'morphology of landscape' as the human habitat, that is accessible to the human senses (*sichtbaren Züge*). This aim is evident in the works of his students such as the early studies of Schott on the settlement pattern of Ontario, Wilhelmy on the rural settlements of Bulgaria and his study of the city of Sofia, all published in the geographical series of the University which was founded by Schmieder.

Schmieder has always been an alert and discriminating observer and recorder of the visible scene and also an ardent user of the library and the document (in English and Spanish, as well as German) in order to interpret the visible scene of the present. Habit and habitat, society and landscape are basic to his approach that so closely resembles that of Sauer in America and Deffontaines in France, and draws so much from Hettner and Schlüter. Schmieder created a school of thought at Kiel, and he founded a series of research monographs in 1932. He was an enthusiastic teacher of the methods and purpose of geographic observation in the field and sought to introduce students at the earliest possible date to the history of geography and to the journeys and observations of the great makers of geography.

The publications of Schmieder culminate in a great work on *Die Alte Welt*, of which the first of two volumes was published in 1965 on the Orient ('steppes and deserts of the nothern hemisphere and their border zones'). This work is in the series of handbooks that was founded by Ratzel and Penck, today edited by Lautensach and Kolb, and the first volume is dedicated to Alfred Hettner.

The big geographies of the Americas, that first appeared in the early thirties, have been rewritten and were published as an entirely new work in two volumes (*Middle and South America, and North America*) in 1962 and 1963. In 1953 there appeared in Mexico a book in Spanish on the Old World. Among Schmieder's research studies, special attention is drawn to the origins of the pampa grassland (1927), the original seat of settlements in Brazil ('the Brazilian culture hearth') 1929, and his essay on the East Bolivian Andes (1926), all published by the University of California (Berkeley). His latest pronouncement, dated 1955, is on 'agrarian systems and field patterns and their influence in shaping the cultural landscape'. This was the presidential address to the Pakistan Science Conference at Karachi. In the work of Schmieder one finds a felicitous combination of the approach of Schlüter, Hettner, and Gradmann, that has resulted in a substantial step forward in geographic research and literary presentation. Unbeknown to Schmieder, I myself found, through his teaching and writings, a guide through the varied threads of theoretical discourse in Germany in the middle thirties.[1]

Further References

An exhaustive listing of publications in the period of the Second World War appears in the series *Naturforschung und Medizin in Deutschland* 1939–46 as part of the FIAT review of German science. Volumes 44, 45, 46 and 47 deal with Geography and are all edited by Hermann von Wissmann of Tübingen. This is an invaluable bibliographical compilation of all aspects of geographical research in Germany during this period.

[1] General research trends in the inter-war period are discussed in N. Krebs' essay in the *Geog. Zeit.*, 44 (1938), pp. 241–77. 'Die geographische Wissenschaft in Deutschland in den Jahren 1933 bis 1945', *Erdkunde*, 1 (1947), pp. 3–48, subsequently translated and published in *Annals. Ass. Am. Geogs.*, 39 (1949), pp. 99–137.

13

Leaders of the Fourth Generation

The purpose of this and the following chapter is to trace the dominant research trends in the last twenty years. This chapter will give attention to three of the leading personalities, whose ideas and substantive works have had repercussions not only on their special fields, but also on geographic scholarship in general and on the public interest. We have selected the following as such leaders—Hermann Lautensach, Carl Troll, and Hans Bobek. Others are omitted from the original manuscript solely owing to lack of space. The following chapter is an appraisal of research trends rather than of particular contributors. Taken together these two chapters reveal the significant problems and prospects of geographic scholarship. The same procedure will be followed later for France.

HERMANN LAUTENSACH

Hermann Lautensach (1886–   ), who reached his eightieth year on September 20th, 1966, is, in the words of Carl Troll, 'the undisputed master of research in regional geography' in Germany.[1] He was born in 1886, the son of a distinguished scholar and gymnasium teacher in Gotha. Here was located the old-established cartographical institute of Justus Perthes and Lautensach as a boy had the opportunity of frequently meeting its director, A. Supan. He also made a close life-long friendship with Hermann Haack, who became one of the world's greatest cartographers at the Perthes Institute until his death in 1966.

Supan advised Lautensach to study with Hermann Wagner at Göttingen, whither he went in 1905. He transferred briefly to Freiburg, and then to Berlin where Albrecht Penck had just succeeded

[1] Carl Troll on Hermann Lautensach in *Erdkunde*, 20 (1966), pp. 243–52. These biographical notes are based on familiarity with the works of each scholar, together with personal statements by letter.

161

Richthofen. Lautensach was an assistant to Penck and in 1910 took his doctorate (*promoviert*) with a work on the glacial morphology of the Ticino, and reviewed in a long article the great work of Penck and Brückner on the Ice Age in the Alps. His life-long association with Penck is reflected in his presenting the memorial speech in 1958 at the master's graveside at Stuttgart on the occasion of Penck's 100th birthday and on the occasion of the meetings of the German Quaternary Congress, a meeting primarily of specialized geologists.

In spite of the stimulating start of his academic life under Penck, Lautensach decided to become a teacher and from 1911 to 1927 (excluding military service in World War I) taught at a gymnasium in Hanover. He was an active editor of the *Geographischer Anzeiger*, the journal for the teaching of geography, together with H. Haack of the Perthes Institute. In 1924 he accepted Karl Haushofer's invitation to participate in the editing of the new periodical *Zeitschrift für Geopolitik* together with two other outstanding geographers, Otto Maull and Erich Obst. But all three, disliking the political trends of the periodical, withdrew from their association well before 1933. In the twenties he revised the three-volume standard work of Supan on 'school-geography', wrote two volumes to accompany the *Stieler Handatlas*, and completely revised the Sydow-Wagner *Methodische Schulatlas*.

In 1927 he left his school post in Hanover to become assistant to Fritz Klute (who was almost the same age) at Giessen. His work for the *habilitation* was on the coastal morphology of Portugal (1928). For this work, he carried out field studies over a period of three years and at this time he published several climatic and geomor-phological studies of the Iberian peninsula, and turned also to problems of cultural geography. He finally produced a compre-hensive work on Portugal, in recognition of which he received an honorary doctorate degree of the University of Coimbra in 1937.

He now turned his skills and enthusiasm to a remote corner of the world, Korea. He traversed 15,000 kms. with 'almost unimagin-able energy and endurance' (Troll). From this there emerged a large standard work (1945), and a second shorter version in 1950.

Lautensach emerged as a leading geographer in Germany on his return from Korea. He was invited to the *Technische Hochschule* in Braunschweig in 1934, but in 1935 he succeeded Gustav Braun at Greifswald. While there he wrote his big studies of Korea over a period of ten years until the end of the Second World War. In 1939 Lautensach, after the death of his first wife, who was Portuguese, married a second time, his wife being a geographer and a co-worker. After the war (1945) already sixty years old, he moved voluntarily, without a post, to West Germany. He found a post in Speyer for a short time, but in 1947 accepted a chair at the *Technische Hochschule* in Stuttgart and remained there until retirement in 1954. This institution was badly damaged by bombing during the war, and was in ruins and had to be completely rebuilt and re-equipped. His researches, however, have continued since retirement and proceed actively in 1965 in his eightieth year.

Lautensach, says Troll, is one of the few living geographers who have researched in fields of both physical and cultural geography, encouraged to this end by his pursuit of regional geography. He has always insisted on clarity of concepts and limits of geography, and follows Krebs, ten years his senior, in this respect. Both were pupils of Penck. It was from Krebs that he took over the editorship of the *Geographischen Handbücher* and he completed Krebs' unfinished work on comparative regional geography (*Vergleichende Länderkunde*), whose third edition was published in 1966. Lautensach's theoretical framework of geography is based on a sound knowledge and appraisal of the natural sciences, mathematics, and philosophy, as well as pedagogy. He is very conversant with non-German thought in geography, for he has fluent command of Portuguese, French, English and Spanish languages. He belongs to three scientific academies and has been awarded five distinguished medals.

Lautensach has devoted much thought to the theory and practice of geography, which he has sought to clarify *vis-à-vis* the physical and social sciences. He wrote on this theme in the *Handbuch* in 1933. He has since developed a theory of regional geography which he first expounded in full in 1951 in a colloquium to the memory of Richthofen, on the opening of the large new institute of geography in Bonn—a highly significant occasion. The areal variations of the earth's surface—which is the essence of the regional concept—are examined in their associations all the way from a local to a worldwide scale—precisely as envisaged by Hettner and Penck. Lautensach arrives inductively at a hierarchy of world units comprised within a major framework of 40 unit areas (*Grossräume*). He has recently applied the theory in a detailed study of the Iberian peninsula, to which we shall return in the next chapter. The *Formenwandellehre* has been Lautensach's major concern for three decades, and, as Troll attests from experience in teaching, is of 'great heuristic and didactic value'. The scheme, discussed below, though applicable in the realm of physical phenomena, has its limitations in the realm of cultural and social phenomena, as Hans Bobek has pointed out. In this realm other categories are to be included, such as cultural and economic associations, but Lautensach's scheme, concludes

Troll, undoubtedly forms an effective springboard for future work in understanding the areal variations of the surface of the earth.

Lautensach, though originally trained under Penck, early veered towards the views of Otto Schlüter, whom he regards primarily as his mentor. 'This reflects the great stimuli that Schlüter, has given to geography since the turn of the century' (Troll). Lautensach wrote an appraisal of Schlüter's work in 1952 in *Petermanns Mitteilungen* to commemorate his 80th birthday. He seeks here to blend the divergent views of Schlüter and Hettner that dominated geographic thought in the first third of this century.

CARL TROLL

Carl Troll (1899–   )[1] professor of geography at the University of Bonn since 1937, and president of the International Union from 1960 to 1964, is a leading world-geographer of the post-war decades.

Born at Wasserburg-am-Inn in Bavaria, Troll entered the University of Munich in 1919 and studied the natural sciences, including botany and geography. He took his doctorate in botany in 1921. In that same year (aged 22) he was engaged by Erich von Drygalski, professor of geography, arctic explorer and pupil of Richthofen, as his personal assistant, on condition that Troll undertake research on a phytogeographic topic (combining geography and botany) for his *habilitation*. Troll writes to me that at this time his development was much influenced by Th. Herzog, the professor of botany at Munich and expert on the vegetation of Bolivia, and Robert Gradmann, whose researches are discussed on pp. 144–47. The *habilitation* thesis was on the influence of maritime conditions on the vegetation of central Europe (1925), a work that triggered off a chain of productive research. His first research on the glaciology of the Alpine Foreland, much influenced by Penck, was published in 1924, beginning around his home town in the Inn-Chiemsee area. The subtitle of this study, 'the geographic aspect of a typical glacier of the Alpine foreland', reveals its essential geographic purpose. This work followed up that of the master, A. Penck. In 1926 appeared a monograph on the young glacial gravels on the periphery of the German Alps—their surface features, vegetation and landscape (*Landschaft*). The field researches of Troll on the glacial morphology of the Bavarian foreland of the Alps, concluded at this stage, around 1930. There is a remarkable parallel in area and purpose in glacial geomorphology to the pioneer work of A. Penck in the Bavarian Alpine

[1] Hermann Lautensach on Carl Troll in *Erdkunde*, 13 (1959), pp. 245–58.

Foreland. Indeed, Troll states in a personal letter that, although he did not study under Penck, he considers himself as one of his followers, and this is abundantly evident in his first field work in the mid-twenties on glacial geomorphology and on the concept of the landscape unit. So ended the first phase of Troll's career.

Troll says that, while working at this time (the 'twenties) on the origins of glacial terraces in his home district of the Inn-Chiemsee area, he became aware of the interdependence of soil, slope and vegetation. Thus, he says, he became a geographer, or, as he prefers to express it, a regional ecologist, in which geomorphology, soil science, plants, climate and hydrology are all interrelated. In 1924 (financed, on the recommendation of Drygalski, and the support of the *Deutsche Forschungsgemeinschaft*) he travelled for three months in northern Europe.

In May 1926 Troll set off for his first overseas research expedition. He went to the tropical Andes, choosing this area through the influence of Th. Herzog and A. Penck. He remained there for three years and pursued both morphological and botanical studies and undertook a stereophotogrammetric survey of the Cordillera as a member of the mountaineering expedition of the *Deutsch–Österreichischen Alpenrereins* (1928). Troll writes: 'I collected 16,000 specimens (all burned during the War), took aerial photographs, studied past glaciation, structural soils, and Indian (native) agriculture'. He travelled by air over Columbia, Ecuador and Panama at the invitation of an airways corporation and thus learned by experience the meaning of aerial-photo interpretation.

He returned to Germany in September, 1929, and moved to Berlin as professor of colonial geography. In the following years—through the thirties—he published many papers on the Andean researches, in which he dealt with geomorphological, cultural and economic matters. In 1930 a paper was published on the 'economic-geographic structure' of tropical South America, the theme of his inaugural lecture as professor in Berlin. Most important among these published field-researches was that on the vegetation of the tropical Andes according to *Landschaftsgürteiln und Landschaftsstruktur* (1931). Throughout his life Troll has been deeply concerned with the relations of vegetation to altitude and climate in the mountains of the tropical Andes.

With his base in Berlin, Troll undertook research travel for one year (September 1933–August 1934) in East Africa, from Eritrea to the Cape, and there examined questions of climate and vegetation and of 'colonial' settlement. He also accompanied an expedition to Nanga Parbat in 1937, and thereafter returned to East Africa. In

Germany

1937 (May–July) he visited Nanga Parbat for a second time as a scientist with a German mountaineering expedition led by K. Wien, and from his observations produced a remarkable map of the vegetation (1:50,000). In the same year he followed a route from Darjeeling across the southern slope of the Himalaya to the Tibetan border, and produced a profile of the vegetation.

At the end of 1937 Troll became professor at Bonn. He continued to work on questions of climatology (especially variations with altitude in the tropics), and produced a new classification of climate and vegetation. In 1944, among many other research studies, he wrote a monograph on 'structural soils, solifluction, and past-climates of the earth'. In 1948 he reported on periglacial phenomena, geo-chronology, and climatic change in Europe as evidenced by pollen analysis.

In addition to his masterly contributions to geomorphology and biogeography, Troll has long been actively engaged in human geography. As professor of colonial geography at Berlin in the thirties, Troll busied himself on the possibilities of white settlement in East Africa. He has recently edited the *Grossen Herder Atlas* (1958), and has also contributed to the methodology of geography, for he was one of the first to appreciate the value of the air photo to geographic study (beginning in 1939 until 1943).

One of Troll's major achievements was the foundation of the periodical *Erdkunde* in 1947. He also made in 1947 a remarkably comprehensive survey of the progress of German geography from 1933 to 1945. This is a scholarly appraisal from one who, during the Nazi period, continued his scientific work and refused to be associated with the national socialist government. A translation of his paper was printed in the *Annals of the Association of American Geographers* in 1949. Troll's views of the purpose of geography are discussed in the first volume of the *Bonner Geographischen Abhandlungen* which he established in 1947.

Troll celebrated the 40th year of his entry into academic life in February 1965. He received an honorary degree at the University of Vienna at its 600th year celebration, and has been awarded honorary membership of scientific societies in Denmark and Italy. He retired in 1966. In the career of Troll, one is repeatedly reminded of two other makers of geography—Alexander von Humboldt, who carried out long field researches in the same area of South America; and his senior contemporary and close associate, Albrecht Penck, whose interests and field of enquiry in glacial geomorphology and regional ecology find remarkable parallels (and acknowledgements) in the works of Carl Troll.

Leaders of the Fourth Generation

HANS BOBEK

Hans Bobek (1903–   )[1] was born at Klagenfurt in Austria but his family moved to Innsbruck in 1909. There he spent his school years, and attended the University of Innsbruck. In his monograph on Innsbruck (1928), the city was interpreted as the spatial expression of the forces of urban life, rather than as an architectural assembly of building structures. This work was undertaken in the geographical institute, which was headed at that time by Johann Sölch, an early Austrian student of Albrecht Penck. Bobek served as an assistant for three years to Sölch, though the latter made little impact on his assistant's outlook. Bobek's emphasis on the appraisal of function as the key to the geographical interpretation of cities lies behind all his subsequent work on the theory of 'social geography', which has had a wide impact among students of the social sciences in Germany.

In 1931 Bobek went to the University of Berlin as a specialist in the urban field at the invitation of N. Krebs. But he turned to geomorphology in order to qualify under Sölch for his *habilitation* and to gain entry to a University career. This research he actively pursued in the Ziller and Inn valleys and it was enthusiastically received by Penck but was hardly what Krebs had foreseen. In December of 1935 he took the *habilitation* in this geomorphological research, but for political reasons he was not permitted to give lectures in Berlin for three years. In 1934, he made his first field journey to Persia and went again in 1936 and 1937.

Persia or Iran became his main land of interest. He turned at first to questions of physical geography such as climatic changes in the Pleistocene and prehistoric periods. He also prepared a map of the Sulaimann mountains on a scale of 1:100,000. Further research into the urban ways of life in the Orient were projected in this area, but these plans were stopped by the outbreak of war in 1939. During the war he was engaged in intelligence work in the Orient and also did work in cartography and photogrammetry, for which he was eminently qualified, and which stood him in good stead after the war.

Bobek was placed at Freiburg-in-B. from 1946 to 1948. In this short period he produced an important series of research papers on the geography of agriculture and settlement in southern Baden. Schlüter tried to attract him to Halle as his successor, but Bobek decided in 1949 to go to the *Hochschule für Welthandel* in Vienna. In 1951 he moved to the University as the successor to Hugo Hassinger.

In Vienna, the capital of his homeland, he became preoccupied

[1] An appraisal by W. Hartke in *Festschrift zum Geburtstag von Hans Bobek*, 1963, pp. 5–22.

with his ecological interests. He examined the vegetation of Persia as the expression of climatic change and destruction by man. Climatic change through historic time and its impact on 'landscape-ecology' in Iran is the theme of a paper in 1955. In two further trips to Iran, in 1956 and 1958–9, he was primarily concerned with questions of social geography and had plans for a work on the regional geography of Iran.

Social geography, since the days of his initial work on Innsbruck, has been an outstanding contribution of Bobek's work. He has reported on various occasions on the purpose of social geography, based on his specific studies in two culture realms — western Germany and the Orient. Especially notable, and accessible now in English (Wagner and Mikesell, *Readings in Cultural Geography*, 1962) is his essay on 'the main stages in socio-economic evolution from a geographic point of view' (1959). The focus of such areal study of human groups lies in the existing geosocial structures (*Lebensformen*), their impact upon each other, on production, and on the modes of grouping of settlement. Social geography is *not* the equivalent of the geography of man, nor is it merely a branch thereof. It is 'the treatment of the human element in the framework of the total geographical scheme'. It is therefore concerned with the origin and impact of 'anthropogenic forces' on social groups as spatial structures (*Lebensformgruppen*). These structures, in the words of Hartke (paraphrasing Bobek), are the 'most essential human entities and are therefore the most essential geographical (regional) units'. This emphasis of social forces, basic to the approach of Bobek and Hartke, is fundamentally opposed to that of E. Otremba who gives predominant emphasis to economic forces. It appears that while the thinking of Otremba runs close to that of P. George in France the ecological approach and emphasis of Bobek (and Hartke) has much in common with that of Max Sorre. The predominance of non-economic forces (that is, social forces), says Bobek, is particularly apparent in the 'under-developed lands'.

14

Post-War Trends

GENERAL[1]

'*Die Geographie ist Raumwissenschaft schlechthin.*' So writes Carl Troll.[2] This means in translation, that geography is basically an areal science, or, as we prefer to state it (and we are sure that Troll would agree), a regional science. This statement, expressed entirely independently by one of the foremost continental scholars, reveals the essential concern and springboard for all geographic work. Alternatively, Hermann Lautensach, the dean of German geographers, considers that geography studies 'the lands of the earth from the genetic, causal and functional points of view'.[3] He defines its conceptual framework more specifically than Troll as follows:
1. To investigate the phenomena on the earth's surface, so as to explain the development of the individual components over the whole of the earth.
2. To investigate the interaction of the components as specific assemblages in particular areas of the earth and to present them in a logical arrangement.

The first, the analytical approach, is the field of general geography. This falls into the traditional fields of physical geography and cultural geography (*Kulturgeographie*). It would be clearer to refer to the latter as *Anthropogeographie* in German usage, *géographie*

[1] W. Hartke, *Denkschrift zur Lage der Geographie, Im Auftrage der Deutschen Forschungsgemeinschaft*, Wiesbaden, 1960.

[2] C. Troll, 'Der Stand der geographischen Wissenschaft und ihre Bedeutung für die Aufgaben der Praxis', *Forschungen und Fortschritte*, 30 (1956), Heft 9. pp. 257–62. See also C. Troll, 'Die geographische Wissenschaft in Deutschland in den Jahren 1933 bis 1945: Eine Kritik und Rechtfertigung', *Erdkunde*, 1 (1947), pp. 3–48.

[3] H. Lautensach, 'Der geographische Formenwandel: Studien zur Landschaftsystematkik', *Colloquium geographicum*, 3 (1952), Bonn, pp. 1–191 and 'Über die Begriffe Typus und Individuum in der geographischen Forschung', *Münchner Geog. Heft* 3, (1953), pp. 1–33.

humaine in French, or human geography in English. The second, the synthetic approach, may be referred to as regional geography (*Länderkunde*). I would add that all geographic work, at the local, continental or world-wide levels, must be conceptually based upon these fundamental objectives. This principle is enunciated again and again by its makers, and gives meaning and purpose to the geographic study of individual places (microcosmic) and world-wide areas (macrocosmic), and clearly brings together 'regional' and 'general' geography.

Systematic or general geography, writes W. Hartke, has allies in the systematic sciences of geology, zoology, botany, and anthropology. But regional geography, as a synthetic study, has few sisters. The most closely related are philosophy and history. In the first half of the nineteenth century the synthetic study of areas held sway, based upon the philosophical concept of the interdependence of phenomena. Out of this emerged the search for a typology of landscape (*Landschaft*) and the individuality of every land (*Individuum*).

Many earlier branches of physical geography have become separate disciplines in the twentieth century. Such are geophysics, geodsy, geomorphology, meteorology, cartography and oceanography. The relation of geography to each of these has developed differently in different countries, but in general, in Germany, these subjects lie in the field of general physical geography, though research in them is also pursued in geophysics, meteorology, botany and pedology.

The trend has not been the same in general cultural geography, continues Hartke. Distinct disciplines have not developed as in physical geography, but they have been shared between geography and ancillary disciplines. Ethnology, that was actively pursued by geographers at the end of the nineteenth century, has now passed over almost entirely to anthropology. The new growth of the economic and social sciences has taken place independently of geography, though new problems and techniques are now bringing them closer together.

The modern field of regional geography (*Länderkunde*) is summarized as follows by Hartke. It uses the contributions of ancillary sciences to understand particular areal complexes. General human geography in Germany has concentrated on the historical development of landscape. This refers particularly to field-systems and patterns of farm settlement as they have developed since the Middle Ages with the disappearance of medieval settlements (*Wüstungeprozess*) and the resettlement of abandoned areas. Similarly, the operation of economic and social forces in an area has been examined in their interaction with physical conditions to produce distinct regional associations (*Landschaften*). Special attention is now being given to the spatial variants of economic and social forces, in their operation in transforming the structure of the landscape. This is a functional as distinct from a genetic approach and characterizes much current work. It had its predecessors in the work of Leo Waibel and Hans Schrepfer and especially A. Rühl. The approach is evident in the study of both agrarian and urban geography. From this trend there emerges a growing interest in the application of geographic techniques to problems of physical planning and areal development.

PHYSICAL GEOGRAPHY[1]

Research trends over the past twenty years in physical geography have been reported to me by Professor F. Wilhelm of the University of Kiel.

Researches in physical geography continue actively in highly specialized aspects and problems, but the institutes of geography at University level seek to maintain the unity of all the specialized branches as one whole. Before the war there was usually only one professor. Today there are at least two and sometimes three and four, and future plans involve an increase in numbers and therefore in specialization. A certain number of separate special departments for particular branches are already in being.

Physical geography, especially geomorphology, is an essential area of study in all geographical institutes in Germany. But in spite of the efforts to maintain the 'unity' of geography, there is, in fact, a very marked specialization in physical as well as in human or regional geography between one University and another, depending above all upon the special interest and qualifications of the professors.

Professor Wilhelm concludes his statement as follows:

This short review makes clear that in spite of the efforts to maintain the unity of geography, at least in research, there has emerged in the last decade a sharp differentiation in various branch disciplines. This trend is an essential consequence of the march of research that obliges narrower specialization.

I would wish to make two comments on this review. They have

---

[1] This section is based on a review, with a full bibliography, by Professor F. Wilhelm of the University of Kiel, dated 10th March, 1965.

been made before, but they need to be again emphasized at this point. The first is that so many outstanding geographers in Germany, as in France, have made, and continue to make, substantial researches in both fields of geography, physical and human. This applies, for example, to the works of H. Mortensen, H. Schmitthenner, H. Wilhelmy, and H. Bobek. But they are all essentially geographers and maintain a composite perspective in spite of their specialized researches. Many still manage to remain ambivalent. The second point is that, as Wilhelm points out, the fields of investigation in geomorphology, climate or vegetation are now so highly specialized that the training and interests in aspects of physical geography are far removed from the training and interests in the field of human geography, so that specialists in each camp are becoming of necessity ever more further removed from each other unless continuing interest is maintained in one particular area of the world.

HUMAN GEOGRAPHY[1]

The morphology of the cultural landscape, as set out by Schlüter in 1906, soon became firmly rooted in German geography. His views have had a profound influence on many geographers, including W. Credner, H. Hassinger, N. Krebs, O. Maull, Fr. Metz, H. Dörries, and L. Waibel. Indeed, the first full exposition of human geography as the 'morphology of landscape' came from Hassinger in the *Handbuch der geographischen Wissenschaft* (Vol. II, 1933–36). This work expressed the viewpoint at that time in the light of work done in the preceding twenty-five years and is generally accepted today.

Research into the characteristics of the original native landscape has been a strong feature of German geography, though (as in France) it has long been the concern of other specialists. This study was most closely associated in the first half of this century with the work of Robert Gradmann on the natural landscape of the Neolithic era (p. 144) and Otto Schlüter on the vegetation of central Europe in the early Christian era.

Recent work seems to agree with Gradmann's general distinction between forested and forest-free areas in the Neolithic, though not with his clear cut correlation of these with later and older settlement. Moreover, it is now argued that the steppe-heath was a plant

[1] H. Overbeck, 'Die Entwicklung der Anthropogeographie (insbesondere in Deutschland) seit der Jahrundertwende und ihre Bedeutung für die Geschicht-liche Landesforschung', *Berichte für Deutsche Landesgeschichte*, (1954), pp. 182–244.

formation between woodland and steppe. This whole matter remains one of scholarly controversy and the fact that geographers have been continuously involved in it is evidence of the importance of their contribution. Continuity of settlement in the prehistoric periods is only true of the core areas of the open land, whereas there seems to have been considerable coming and going in the peripheral areas. The limits of the woodland shifted between the oldest Neolithic period and the early middle ages in the areas of Slav settlement. There were forest clearings in the Iron Age period in various wooded uplands of central Germany and there was a recession of settlement in the post-Roman period in the west and south of Germany. Gradmann underestimated the ability of prehistoric man to clear trees in forested land.

The study of the landscape has been furthered by examination of changes since the eighteenth century as a result of the great developments in agrarian structure and in the techniques of industry and commerce. Three other periods of change are the period of the medieval forest clearings; the rise of towns in the Middle Ages; and the period of the territorial lords and their mercantilist economy in the seventeenth and eighteenth centuries. Much attention is given in this approach to the morphology of town and village, not only to description but also to morphogenesis. Research studies include changing aspects of urban and rural economy and the development of urban and rural house types as culture forms, that crystalize in particular areas of characterization and spread from them.

Recent German work emphasises a functional or ecological approach. Social geography—the study of human societies as spatial groups—has become an integrating segment of geographical science. This has led to further study of functional complexes or spatial entities (*Raumeinheiten*) as distinct from, though related to, landscape complexes. But we should not be misled into believing that this is an antithesis or a rejection of the morphological approach. Schlüter did not wish to exclude religion, ethnography, speech and State from geographical study, but sought to limit such study to an appreciation of what is evident in the landscape. More recently, in the fifties, H. Bobek and W. Hartke have emphasized the anthro-pogeographic *Kräftelehre* and their territorial expression as 'anthro-pogenic living areas' or 'functional areas or entities'. Leo Waibel, who followed Schlüter, warned against a one-sided physiognomic approach and urged the approach of Hettner, who emphasized the inner character or personality of an area (*Wesensheitslehre*). Waibel wrote that 'the character of the cultural landscape depends not merely on its aspect, but also on its social, economic, legal and spiritual structure and not least upon its place in history' (1933).

Function means, in this usage, the spatial interconnections of phenomena and their arrangement and development through time. Its main contributors have been H. Bobek, W. Christaller, W. Hartke, Th. Kraus, W. Müller-Wille, and the late H. Schrepfer and L. Waibel. It is on these lines that the field of 'social geography' has been recently developed in Germany. Societies in their spatial variations are examined, together with the changes they induce in the aspect of the landscape. Thus the landscape has a double aspect, the natural plan (*Naturplan*) and the cultural formative plan (*Gestaltplan*), behind which lie the forces of the social-geographical system. There are in fact within geography two kinds of unit, one physiognomic or morphological, the other physiological or functional. Indeed, *Landschaft* in German usage means not only the visible landscape but also the socio-historical unit in which there occurs an association of related landscape elements. Functional units of a wider order are the tributary area of a town, an economic complex, a cultural complex, or a political complex. All these are anthropogenic *Lebensraüme*. They are what Hettner sought for in his emphasis on *Wesen*, over and above Schlüter's *Bild*. This approach is clearly adumbrated by Hassinger in the *Handbuch der Geographischen Wissenschaft* in 1933.

Geographers turned early to this concept of the functional entity, and sought to use this approach to define the units that should be used in a new political framework. It is concerned essentially with the analysis of spatial systems, in terms of nets and nodes, core areas and their boundaries. Currently, particularly in American circles, one speaks of ecosystems. Such studies in France and Germany go back over fifty years. This is evident in the thirties in the examination of the forms of agricultural occupance by Waibel and in Christaller's concept of the central place and its field of association. Von Thünen's theory of the zoning of land uses with distance around a central place was developed by Waibel in the thirties in his interpretation of agricultural distributions. On this basis also Müller-Wille built up what he called a 'functional chorology' of Westphalia. Political entities have also been examined as *Lebensräume*, that fall into a hierarchical system, and this mode of attack has become sharper in the post-war years. The social and psychological factors involved in the establishment of functional entities have been emphasized by E. Otremba, W. Hartke, and H. Bobek. Hartke has used newspaper circulation as an indicator of community of ideas or attitudes (an approach incidently that was used by the Burgess school in Chicago over thirty years ago). These associations can often be understood only in terms of their formation through time, and here the cultural historians and human geographers in Germany find much common ground.

THE REGIONAL CONCEPT

The regional concept[1] examines the tendency for phenomena—physical, biotic and human—to associate in complexes over particular areas of the earth's surface in such a way as to give distinctive character to such areas from their surroundings. Such an association is a spatial entity in respect to its visible landscapes, the interraction of its associated phenomena, as well as in its spatial relations with surrounding areas. Its central concern is not with what Hartshorne has criticized as the 'phenomenology of landscape', such as the depth, length, and structure of bridges, the (obscure) origins of a particular architectural style of building to be found in farmsteads, the methods of construction of walls and fences, or the occurrence and movements of particular atmospheric disturbances; nor is its central concern with the distribution of these or other phenomena, such as individual plants or arrows among primitive peoples. These data are essential for understanding terrestrial areas and the geographer may be obliged to pursue them if not already available—this indeed is one of the main ways in which he has stimulated research among other scholars who have more competence to do so than he himself—but the geographer's central concern is with the tendency for such phenomena to be segregated in complexes which have a locale, a core, and a more or less clearly defined gradient of change on their periphery. This is what is meant by the areal associations of phenomena and this is how Hettner logically assessed the relevance of any particular phenomenon or distribution to geographical study.[2]

[1] H. Lautensach, 'Otto Schlüter's Bedeutung für die methodische. Entwicklung der Geographie: Ein Kritischer Querschnitt durch ein Halbjahrhundert erdkundlicher Problem-stellung in Deutschland.' *Pet. Mitt.*, 96 (1952), pp. 219–31. In addition to the works of Lautensach, see the comprehensive appraisals by E. Neef, 'Zur grossmasstäbigen Landschaftsökologischen Forschung', *Pet. Mitt.*, 108 (1964), 1/2 Heft, pp. 1–7; and G. Haase, 'Landschaftsökologische Detailuntersuchung und naturräumliche Gliederung, *ibid*, pp. 8–30.

[2] There is a formidable bibliography of substantive research and of methodological discussion and critical evaluation of trends. I have selected representative statements from recognized authorities as indicative of trends in Germany. For a reliable appraisal of the development of the regional concept and its status in the thirties see K. Burger, Der Landschaftsbegriff: Ein Beitrag zur geographischen Erdraumauffassung, *Dresdner Geog. Studien*, Heft 7, (1935). In the following pages the views of Lautensach and Troll are closely followed.

Germany

A review of this concept was published some years ago by H. Lautensach.[1] The concept of a region, he concludes, has two meanings. It is both an individual or specific unit area, and typological or generic area. It is unfortunate that *Landschaft* and landscape are basic words in two languages and it is a pity, says Lautensach, that Sölch's *chore* has not found acceptance. The difference between the concepts of individual and typological, of specific and generic regions, is, in his opinion, one of the great stumbling blocks in the way of geographical research.

The relation of the geography of societies to that of landscape continues as a central professional question. If geography is not to burst asunder, says Lautensach, it must retain its focus on landscape, as the human environment. The geographical study of societies, independent thereof, is a subsidiary interest and might well be relinquished as an end in itself. Hassinger refers to this as *Kulturgeographie*, and to the wider study of human societies as *Soziogeographie*. To this latter he gives a negligible portion, 48 pages, of his 376 pages in the *Handbuch*. Hassinger writes in his essay in the *Handbuch*: 'We speak today of the geography of dialect, language, literature and art . . . such researches have only the spatial method in common with geography, not the objects.' The days of geography as the study of distributions is long past. It is of interest to compare these views with those of the French geographers, Brunhes and Vallaux as discussed below on pp. 212–14.

The development of the field of social geography, says Lautensach, shows the influence of Schlüter. Van Vuuren in Utrecht recognized (in 1941) the role of society in its area of occupance in creating a distinct kind of landscape[2]; this is again reflected in the current ideas of Th. Kraus and H. Bobek.[3] The forces changing man's occupance of the land are group forces. The individual belongs to a variety of such groups of forces or ideas in areas of advanced society. Thus, there is a variety of social structures each operating over a distinct 'social area'. Among these group

[1] H. Lautensach on Otto Schlüter etc. in *Pet. Mitt.*, 1952, is a reliable appraisal on trends of the regional concept in Germany over the last generation or more.

[2] Van Vuuren, 'Warum Sozialgeographie', *Zeit. d. Ges. f. Erdk. in Berlin*, 1941. pp. 269–79.

[3] Th. Krauss, 'Geographie unter besonderer Berücksichtigung der Wirtschafts und Sozialgeographie', *Augfaben deutscher Forschung*, Cologne, 1957; H. Bobek, 'Gedanken uber das logische System der Geographie', *Mitt. d. Geog. Ges. Wien*, 99, (1957), Heft II/III. pp. 122–46. See also H. Schmitthenner, 'Zum Problem der allgemeinen Geographie und der Länderkunde', *Münchner Geog. Hefte*, Heft 4 (1954) and 'Studien zur Lehre von geographischen Formenwandel', *ibid*, Heft 7 (1954).

structures (*Gesellschaftsstrukturen*) are those concerned with the use of land. *These* together form a distinct living-area, which is adjusted partly to the natural features of the land and partly to the pre-existing cultural landscape. Bobek calls this social ecology. Social geography is thus the geography of social areas and their spatial structures. It is based upon a synthesis of the various branches of the geography of human groups and will be called upon gradually to take this place. A distinct contribution to this approach is offered by H. Schmitthenner's *Lebensraüme in Kampf der Kulturen.*[1] The more detailed analysis of this broadly sketched approach is the task of social geography for the future. Such study will surely provide a framework in which to understand the processes of change in the cultural landscape. Lautensach writes:

> Every type of social structure produces, by its impact on the nature of the land, a particular style of cultural landscape, which is expressed in economy, traffic (*Verkehr*) and settlement. These will lead in the future to a departure from the normal division of cultural geography into economic, traffic and settlement geography. This will be displaced by the types of cultural landscape, which emerge from old landscape settings through the operation of the forces of changing social structures.

The finest examples of the development of the regional concept in the last few decades—in the sense of determining and explaining the co-variants of areal distributions on the surface of the earth—are to be found in the works of Hermann Lautensach. His theory has already been noted. It is applied in a major work on the Iberian Peninsula—covering 700 pages and an atlas section of 75 maps—published in 1964.

Lautensach has been working in the Iberian peninsula for nearly four decades and has spent fifteen periods of work in the field. The bibliographic work is reflected in the author's personal card index of 7,000 references.

The first part of the book is a topical analysis of the peninsula as a whole, the second half deals with regional description. In the first half, writes Parsons,[2] 'it is Lautensach the physiographer and climatologist and Lautensach the cultural historian that come especially to the fore, yet contemporary economy is also given its

---

[1] H. Schmitthenner, *Lebensräume im Kampf der Kulturen*, Leipzig, 1938. See also *Lebensraumfragen Europäischer Völker, Band I, Europa, Band II Europas Koloniale Ergänzungsraume*, Leipzig, 1941. A third volume, edited by O. Schmieder, *Gegenwartsprobleme Nord Amerika*, appeared in 1943.

[2] See the review by J. Parsons, *Geog. Rev.*, 56 (April, 1966), pp. 306-7.

due'. He considers historical geography from pre-historic times, physical aspects, human aspects, such as house types, place names, settlements, agricultural systems, and river regimes. In the second half of the book, Lautensach recognizes 27 regions, though he does not use one consistent criterion of definition. 'Rather the glove is made to fit the hand', to quote Parsons. The work, continues the same reviewer, 'approaches the encyclopedic'.

Lautensach's regional treatment is based on the model of 'landscape systematics' that he has expounded and applied for several decades.[1] The whole peninsula is divided into individual units (*Landschaftsformeln*) according to the degrees of correspondence of the geographical facts (i.e., areal variables), and a shorthand is used to designate the common characteristics of each unit. This elaborate notation is essentially the same as Granö's study of Finland a generation ago.

This approach, says Parsons, is a 'confirmation of the unity and traditional character of geography as the study of the earth from the point of view of natural history'. There is no attempt, he says regretfully, consistently and precisely to evaluate 'man's experience as the occupant and modifier of the earth'. The reason for this is clear. Lautensach's purpose is to characterize and explain the variations of the earth's surface in the Iberian peninsula, not to confine himself to the evaluation of man's mode of occupance of this land. These are two different approaches. They are, and always have been, different avenues, with different themes and problems, and different presentations in the field of geographic scholarship.

There is a scepticism, particularly in America, about the regional concept. Yet, in the same number of the *Review* as Parsons' commentary there is an appraisal of a German *Länderkunde* of lower California[2] on the Hettnerian pattern that is essentially akin to that of Lautensach, although it is based on much briefer field work. The work is described by the reviewer as thorough, but condemned as a 'pedestrian compilation', and yet it is lauded as the only study of this particular area, which must be a first priority in the reading list of students of that area. This is overt recognition of the essential need for such 'geographic description'. One suspects that there is an appalling dearth of such studies in the American literature because none have the patience to work consistently in the same area over a long period of time with a strongly focussed sense of problem or purpose. Exceptions are the recent studies of Gottman on *Megalopolis* and

[1] See footnote 3, p. 169.

[2] This is a review of a regional study of lower California in *Geog. Rev.*, 56 (April 1966), pp. 303.

Virginia; but Gottman is a French geographer and has presented to the English speaking world the purpose and method of the geographic scholarship of his country.

To complain of the 'inadequacy of the regional concept', or of the 'quagmire of 'eclecticism' to which it leads, is to condemn the essential purpose and problem of geographic research. The regional concept is the focal concept of all geographic work, and throughout the past seventy-five years has changed in its objectives and methods. It is in sore need, as P. E. James has cogently reminded us, of new understanding and interpretation. 'Regional Geography' is generally regarded as systematic compilation of spatial distributions in an individual area. The widespread scepticism among British and American geographers means that they are, in effect, rejecting or ignoring, the best offerings of their birthright. We have much to learn from the words and works of the past masters. It is for this reason that this book has been written. Lautensach, like many of his German co-workers in the post-war years, is concerned with methods of measuring and mapping the regional differentiation of the surface phenomena of the earth. The procedure has been the keynote to a main productive trend in post-war Germany. To regret that Lautensach does not select and constantly evaluate man-land relationships is indeed to urge the pursuit of a vitally important ecological theme of geographic research. This, however, is not the purpose of Lautensach's work, for he is searching for the areal differentiations of the earth's surface. These are two of the main and distinct themes and problems handed to us by our heritage from the masters of modern geography.

REGIONAL ECOLOGY

Certain aspects of the regional concept have been the subject of much research among German geographers in recent years and constitute such a distinctive feature of their work that we shall give special attention to them.[1] Carl Troll has written on this theme as follows:

It is in the spirit of the times that over the last three decades there has been a strong trend in geographical study towards (a new kind

[1] C. Troll, 'Die geographische Landschaft und ihre Erforschung', *Studium Generale*, 3 (1950), Heft 4/5, pp. 163–81. Also 'Methoden der Luftbildforschung', *Siz. Ber. europaischer Geographen in Würzburg*, 1942–3. Included in a collection of his papers entitled '*Okologische Landschaftsforschung und Vergleichande Hochgebirgsforschung, Erdkunde Wissen, Schriften Reihe fur Forschung und Praxis*, Heft 11 (1966).

of) synthesis. This aims at the evaluation of the spatial inter-connections of the forms of the earth's surface in terms of regional integration rather than in reference to any one set of phenomena. The concern of geography, in other words, is not the mechanics of glacial movement or particular floristic distributions, or the range of particular items of international trade. It seeks to understand the areal interrelation of physical and human spatially-arranged phenomena in terms of the concept of the region (*Landschaft*).

The *Gestalt* and *Ganzheit* philosophy has long influenced the regional concept. It certainly affected the work of Carl Ritter, and has been more definitely worked out in this century by psychologists and biologists. Geographers, however, do not regard regional entities as organisms in the biological sense. Just as life-forms exist as communities (*Lebensgemeinschaften* or *Biozonosen*), so human groups such as family and folk, have a *Gestalt* character as geo-graphical or spatial entities.

The term *Landschaftsgeographie* was coined by Passarge in 1913 but he was not clear as to what this meant, especially vis-à-vis the concept of *Land*. The German word *Landschaft* or landscape is over a thousand years old. Geography has taken it over to express a scientific concept. The geographer refers to a unit area as a *Landschaft* or region, when it has a distinctive physiognomy and a distinctive assembly of 'spatially arranged things'. This has a physiognomic (formal) and a physiological (functional) aspect. The latter is de-scribed alternatively by Troll as functional or ecological.

The geographical region (*Landschaft*) is defined by Troll in the same source[1] as an entity of the earth's surface. It is a *natürliche Landschaft* in terms of its build, the interplay of its spatially dis-tributed phenomena, and according to its inner and outer spatial relations. *Länder* on the other hand, he continues, have admini-strative or political limits, determined by historical territorial divisions or the distribution and limits of particular peoples. 'Natural' (*natürlich*) is used in contrast to 'artificial' or man-made. To avoid confusion, Waibel in 1938 suggested the substitution of 'geographical region' for 'natural region' in the above sense. The term 'natural' is thus confined to physical phenomena, and a *Naturraum* is a unit defined on the basis of physical criteria. A *Kulturraum* is a unit defined on the basis of cultural (human) criteria.

The structure and hierarchy of regional units is Troll's next theme. At this point Troll limits his treatment strictly to the physical criteria and, therefore, to physical units. These fall into a hierarchical series.

[1] Ibid

Post-War Trends

The largest world-wide (generic) units were referred to over fifty years ago by S. Passarge as regional girdles (*Landschaftsgürtel*). The same theme was developed in Herbertson's 'natural regions' in 1905. These systems were based on the world-wide distribution of a few sets of world-wide criteria. They were both based on deductive rather than inductive reasoning. Moreover, such world-wide typology was based on very inadequate data and, therefore, on intelligent guesswork. Troll is concerned with the recognition and understanding of the minor unit areas at the other end of the scale. This is presumably comparable with Herbertson's vague notion of 'tissues and cells', which J. F. Unstead in Britain tried later to develop, using terms such as stow and tract, and D. Linton (from the standpoint of land-forms only) still more recently sought methodologically to clarify. In America, James, Finch, Hudson, Platt and others gave attention to this mode of approach in the thirties. This field of enquiry, however, has found very little response in either Britain or France. It has become, since the war, one of the major preoccupations of many German geographers, and in fact, the recognition and explanation of a hierarchy of regional units, as we have seen, has been acknowledged as the core concern of geography in Germany for nearly two hundred years, as expressed by writers at the end of the eighteenth century and by Ritter and his writings in the first half of the nineteenth century.

What is the dimension of the smallest entity of landscape in this hierarchy? Troll suggests that the smaller the units in the hierarchy, the more important are soils as indicators, as opposed to climate at the highest and widest world level. The size of the smallest unit area of landscape elements depends on the area in which one is located. With the aid of topographic, geological and other maps, plus serial photographs, Troll studied a part of the Rhine Plateau east of the Rhine (1945).[1] The smallest unit area is reached when in 'further breakdown the areal extent of individual landscape elements (*Bausteine*) do not occur in combination with other areal elements, but as individual distributions'. In this smallest unit area there is a recognizable association that forms a unit of landscape (*Landschaftsindividuen*). This is the smallest areal association (*Raumgebilde*). To such a unit area Troll gave the name of 'ecotype', taking it from the biological concept of a 'biotype'. Similar terms used in the German literature are *Landschaftszelle*, *Microraum*, *Kleinraum*, *Naturcomplex*, *Physiochore*, *Kulturchore*, *and Geochore*.

This use of the term 'ecological' apparently refers exclusively to

[1] C. Troll, 'Methoden der Luftbildforschung', *ibid*. See most recently the work of Neef and Haase, footnote 2, p. 175.

the areal coincidence and interdependence of natural or physical conditions, and excludes human works. Such divisions are of basic utility not only to geographers, but also to foresters, soil scientists, plant ecologists, agriculturists, hydrologists and above all, to land planners.

J. Schmithüsen has suggested a distinction in name between the natural (inorganic) unit, for which he proposes the term *Fliese* (*Steinplatte*): and the unit area defined on the basis of the life-forms of vegetation or human uses, to which he reserves the term 'ecotope'. Troll, critically examining this proposed distinction, compares it to the concepts of 'site' and 'cover forms' that the Americans (such as P. E. James) were discussing at this time. This matter triggered off much work through the fifties.

Special mention should be made of the detailed study of the hierarchy of landscape units in various sections of Germany and to the series of maps on a scale of 1:200,000 that is being prepared by the *Bundesamt für Landeskunde* to show the natural areas (*Naturräume*). The latter are compact areas defined in terms of abstracted natural (physical) criteria alone. 'Regional ecology' examines these areal associations of natural phenomena on the earth's surface in their association with life-forms. This system of relationship was defined in the thirties by the British plant ecologist, A. G. Tansley as an *ecosystem*. It embraces land, soil, drainage, climate, and the vegetation and animal forms areally associated with the natural complex. Vegetation occupies the central place in the ecosystem. If the association is left undisturbed a series of successions will take place with regeneration resulting in a climax. Much work has been done in seeking criteria for the mapping of unit areas and their associations. However, we need to go much further, says Troll, in understanding the processes involved in the interrelations of the factors that result in particular areal combinations.

The association of cultural and economic regions (*Landschaften*) now receives attention from Troll. The discussion so far is concerned with *physical* units and *physical* criteria and processes. Human occupance of the land is based on social processes. Human unit areas also occur in a series. Troll writes:

> In the areal (spatial) interconnection of the patterns of Nature and Man in the large and in detail lies the key to an understanding of the regional structure of the earth.

The world-wide mapping of *Kulturlandschaften* has been undertaken many times and will continue to be an essential geographic task. Jaeger in 1943,[1] in preparing such a world classification, used three factors: (1) fourteen major cultural realms; (2) forms of land use (plough, garden, rice, plantation, oasis, shifting hoe cultivation, nomadism, sedentary farming, mining and industry; (3) density of population as an indicator of the degree to which the land has been transformed by human occupance.

We need to turn attention to similar study of small areas, says Troll. He studied in the field, and with aerial photographs, a section in the area of the Rhine Plateau. He found that each (physical) 'site' has a corresponding kind of land use and the nature and process of this adjustment changes through time. This is a harmonic (ecological) relationship (*harmonische Kulturlandschaft*), but more often social and economic factors result in land occupance that does *not* coincide with the physical site conditions. This results in a disharmony in the regional structure. The essential question is 'what forces lie behind these differentials between the human occupance and the site?'. The basic geographical question is to investigate the areal range of human activities and processes, and to determine their impact on the use of the land. This leads to functional analysis and the investigation of functional areas. Even in world-wide patterns of economic regions the spatial interplay of natural and cultural forces is evident as in the ecotope or small area. The geographer in Germany turns to the social and economic forces that lie behind the areal association of human use of the land. Not only are natural units being mapped. We find works by geographers on the impact of urbanization on land uses; the grading and orbits of influence of towns; the forms of agricultural occupance and the current changes in their patterns; and the classification and distribution of modes of human settlement. The functional analysis of this build and the interaction of the forces that create the regional entities is the central concern of scientific geography of this day.

The reader may object that an article written in 1950 is hardly a valid indicator of research trends regarding the regional concept thereafter. On the contrary, Troll's exposition has been very deliberately chosen as a summary and forerunner of subsequent research in this field. It is, however, probably correct to say that the work has been carried out mainly in Troll's institute at Bonn and also by the *Bundesamt für Landeskunde* now in Bad Godesberg. Research has taken place in other directions, and of special note is the following. The attraction of well paid and regular jobs in the cities has caused

---

[1] F. Jaeger, '*Neuer Versuch einer anthropogeographischen Gliederung der Erdoberfläche*', *Pet. Mitt.*, 89 (1943), pp. 313–23. This followed an earlier suggestion in *Pet. Mitt.* 80 (1934).

many country and village dwellers to allow their inherited scraps of land to lie idle (*Sozialbrache*), and the State subsidizes the planting of young trees on private land thus adding another changing element to the landscape. Understanding of the spatial extent of social conditions is essential, in other words, to explain the changing patterns of regional associations.

The purpose of this section has been to refer mainly to work done (in much detail) in Western Germany. The essential thing to grasp is the conceptual framework and hence the distinctive objectives that underlie such studies in the realm of the spatial operation of natural processes, the spatial operation of human processes (social and economic, genetic and functional), and the ways in which these processes are jointly expressed in the areal variations of the face of the earth.

Regional geography or *Länderkunde*, is the core of geographical work. This has been said many times and has recently been urgently reiterated by Oskar Schmieder.[1] This must be recognized and pursued, he urges, if geography is to maintain its identity. The current trend, says Schmieder, is to concentrate on general geography (*Allgemeine Geographie*) and, while Schmieder points this out in Germany, it is even more true in the United States. General geography is supposed to be more scientific and acceptable. This current trend in Germany Schmieder attributes partly to the lack of financial support for work outside Germany. He puts it down mainly to the common plea that one person cannot master the various fields necessary to tackle the synthetic approach that is the essence of *Länderkunde*. This Schmieder will not accept, and attributes it to the great increase in number of professional geographers without a corresponding increase in quality or dedication. He attributes it also to the lack of response to a universal need for attention to practical problems of society, particularly in remote underdeveloped parts of the world that are crying out for synthetic evaluations of their territories. He reminds us that it is almost exactly fifty years (1924) since Albrecht Penck claimed that the carrying capacity of the earth was the most urgent problem for the geography of man, and such a research programme figured as one of his last team projects (see p. 105). This problem is more urgent than ever and calls for the canalization of efforts of geographers in these meaningful directions.

The tradition of *Länderkunde* is as strong in Germany as it is in France. It includes the early works of Humboldt, Ritter, Gradmann,

[1] Oskar Schmeider, 'Die Deutsche Geographie in der Welt von Heute', *Geog. Zeit.*, 54 (1966), Heft 3, pp. 207–22.

Schlüter, Fischer, and Kirchhoff. In our day it includes the works of Waibel, Credner, Lautensach and Schmieder. There are many evidences from publications over the past five years that German scholars are dedicated to and making meaningful contributions to this field, which, as G. Pfeifer declares, repeating almost verbatim the words of Hettner, is 'die Krönung unserer Wissenschaft' (*Geog. Zeit.*, 1964, pp. 329–34). Among these publications are *La Plata Länder* by H. Wilhelmy and W. Rohmeder (1963), *Ceylon* by A. Sievers (1964), *Ostasien* by A. Kolb (1963), the three volumes promised by F. Bartz on the fishing areas of the world (*Die Grossen Fischerräume der Welt*, [two volumes published 1964–5], and the third in preparation) and the two-volume work by Oskar Schmieder on *Die Alte Welt* (1965). The last volumes, it may be noted, are in the *Bibliothek Geographischer Handbücher* that was founded by F. Ratzel and A. Penck and is now edited by H. Lautensach and A. Kolb. Moreover, Schmieder's work is dedicated to Alfred Hettner. The message is clear. The study of particular areas is the core of geography. The challenge is to find distinctive and worthwhile problems and methods of investigation that add up to more than the simple summary of the contributions of the ancillary sciences.

Part Three

FRANCE

15

Introduction

France shared in the intellectual ferment of the late eighteenth century and Paris was an outstanding centre in which many great scholars assembled, with whom Humboldt freely associated for some twenty years after his return from Central America. Several trends may be briefly noted. Remarkable advances in cartography were made in the eighteenth century, especially in the mapping of the world and in geodetic measurement, and the geodetic survey of France was completed in the 1890's on a scale of 1:80,000, so that base maps were available for geographer's research in the ensuing decades. A cartographer to the King, Philippe Buache postulated in the mid-eighteenth century that the earth could be systematically divided into a hierarchy of units on the basis of drainage areas and their surrounding watersheds—a speculation based on the slenderest geographical knowledge, for the greater part of the continental land masses were both unexplored and unmapped. This suggestion of 'natural' units, distinct from the imposed and changeable political units initiated discussions among French and German scholars over the next hundred years, that today, with accurate knowledge and maps of the whole surface of the earth, remains a basic responsibility of the geographer. Comprehensive taxonomic classifications (e.g. of plants, animals and men) and philosophical speculations of human behaviour and history also appeared in the eighteenth century, and none was more significant than Count Buffon's enormous survey of 'natural history' (pp. 11–14).

The really significant development in human knowledge, in its relevance to the developments of modern geography, took place in the third quarter of the nineteenth century with the impact of Darwinian evolutionary thought (after 1859). I refer to the rapid advances in the study of the physical earth and the accumulation and mapping of physical data of land, sea and air, and specifically to the work that was undertaken in France; secondly, to the big advance in the factual study of human societies initiated by Frederic le Play; and, thirdly, to the great work of compilation of Elisée Reclus in a systematic description of the whole surface of the earth, still mainly in the spirit of Ritter and only very partially affected by the march of evolutionary thought and investigation. It was out of these trends and their background that in the last quarter of the century there emerged the new school of modern geographers in France under the leadership of Vidal de la Blache.

Geography in France in the latter half of the nineteenth century was in the hands of a few historians, who simply used the physical environment as a background to history. The notion of the physical unit as a geological entity was put forward by Duffrenoy and Élie de Beaumont in their *Explication de la Carte geologique de la France* in 1841. But their ideas were not furthered by University professors, who merely taught geography as a background to history. Further progress was dependent on geologists, such as De Lapparent and De Margerie. Moreover, there was a division between the Faculties of Science and Letters, so that geographers who were in the Faculty of Letters were not allowed to give lectures on physical geography which was reserved to the Faculty of Science. It was a primary task of Vidal de la Blache to effect a *rapprochement* between the approach of the geologist and that of the historian. Out of this emerged the new geography.

The position around 1900 may be summarized. There had been a chair in geography in the University of Paris since 1809, and this became vacant with the death of A. Himly, primarily a historian, who, like his contemporaries in Germany, made no impact on the development of modern geography. A second chair in Colonial Geography was founded in 1892 and was held by M. Dubois (co-founder with Vidal de la Blache of the *Annales de Géographie*), who was succeeded by A. Bernard. Institutes were established at Lille in 1893, and others followed shortly after at Lyons, Nancy and Rennes. In 1900 Vidal was installed at the Sorbonne, De Martonne was at Rennes, P. Camena d'Almeida at Bordeaux, Lespagnol at Lyons, Bernard at Algiers, and Brunhes at Fribourg (Switzerland). A so-called 'historical geography'—that is, superficial physical geography presented as a background to history, the history of exploration, place names, and territorial divisions—was taught by historians at Besançon, Clermont Ferrand and the *Collège de France* in Paris. This was the situation when Vidal, after a long period in the *École Normale Supérieure*, transferred to the chair of geography in 1899, in succession to Himly. In 1906 there were 27 professors and lecturers in the Universities. After the First World War, geography was represented in all the Universities and nearly all the professors were pupils of Vidal de la Blache. There were then 23 professors as compared with 71 in 1957.

Two special developments call for particular comment. Firstly, political geography has been brilliantly promoted by André Siegfried and Jean Ancel. Ancel, who was a pupil of Demangeon and wrote his doctorate thesis on Macedonia and its contemporary evolution (1930), produced studies on the political geography of the Balkans and Central Europe. He was opposed to Nazi geopolitics and died in a concentration camp during World War II. Siegfried (1875–1959) for long occupied a research chair in Political and Economic Geography at the *Collège de France*. His writings brought him worldwide recognition, and he was elected to membership of the *Acadèmie Française* in 1945.

Secondly, colonial geography has developed since the establishment of the first chair in Paris in 1892. A second chair was founded in Paris in 1937 and was held until his death in 1963 by Charles Robequain. In 1946 new chairs in this field were founded at Strasbourg, Aix-en-Provence, and Bordeaux. In 1947 two further chairs were established at the staff college (*Ecole de France d'Outre Mer*), founded in 1889. The *Collège de France* has a research chair in tropical geography, held by P. Gourou, and there was a well established chair in Algiers, held in succession by Hardy, Bernard and Gautier. The *Atlas des Colonies Françaises* is a superb atlas of the colonial lands of France and its text includes contributions from A. Bernard, E. F. Gautier, Ch. Robequain, J. Weulersse, and P. Gourou, all of whom are professional geographers. These writers, among others, are responsible for many research publications over the past seventy years in North Africa, South-east Asia and the Middle East, and their contributions stand alongside the many monographic studies in various sections of France, which will be discussed on later pages.

It was the task of Vidal de la Blache in formulating the field of geography to reconcile the contributions of his senior contemporaries in France in the fields of geology and history. This he achieved through the interdigitation of land and people within the framework of small areas in France that were accessible on foot, well known at first hand, well mapped, and for which statistical and documentary materials were available.

Physical geography, especially the study of the forms of the land, was advanced by several distinguished scholars in the last decades of the nineteenth century. Dufrenoy and Élie de Beaumont, as noted above, put forward in 1841 the idea of a physical land unit such as the Paris Basin or the Central Massif, as based upon the occurrence of similar geological (lithological) conditions.

> The geological lines which determine the form of the rocks define, as it were, the skeleton of a country, while the hydrographic lines only represent its purely external traits, which, on the same surface, change with time. Moreover river valleys are only isolated furrows, whereas the general modelling of the relief of the earth is limited with geological features.

This statement made by Dufrenoy and Élie de Beaumont in 1841 (and quoted by Gallois) rejected the idea of Buache but it was a generation ahead of its time. Their work preceded the genetic approach that permeated the work of the next generation. Later G. de la Noë and Em. de Margerie, on the other hand, appraised the part played by river action in the development of land-forms (1886). This new genetic approach, recognizing the roles of the agents of erosion and of tectonics and structure in the development of land-forms, was incorporated in De Lapparent's famous *Leçons de géographie physique* (1886). The structural and tectonic themes were taken up by De Margerie in his translation, with many amendments, of the classic work by the Viennese geologist, Ed. Suess, on the 'face of the earth' (1883–1901). Emm. de Martonne, the geographer, produced a masterly appraisal of the whole field of physical geography (land, water, air, plants and animals) in 1909, based on studies throughout the last decade of the nineteenth century; this is a work that has passed through several augmented editions and is still a basic text. The views of W. M. Davis made a widespread impact before World War I in France as well as in Germany, though in France they were received much more sympathetically, and evaluated and modified in field studies by such men as De Martonne, H. Baulig, and Demangeon, rather than meeting with harsh criticism, as was the case in Germany from Hettner and Passarge.

The geography of man—*la géographie humaine*—received critical evaluation by Vidal de la Blache and his colleagues and stood up effectively to the attacks of other French scholars, particularly sociologists, led by Emile Durkheim. The works of Ratzel on anthropogeography and political geography were subjected to searching critical analysis in the pages of the *Annales de Géographie* in the 1890's and 1900's. Historians and sociologists shared in this evaluation and in particular vigorously disputed the ambitious claims and implications of Ratzel's conception of human geography, and the role of 'geographical' (*qua* physical) influences in human history. Ratzel was used as a springboard in the thinking of social scientists in general not only of geographers. Vidal de la Blache and his colleagues developed from Ratzel's 'geographical determinism' what began to be called 'geographical possibilism'. The field of human geography, as formulated by Vidal de la Blache, was fully discussed and its limitations pointed out in an outstanding work by Lucien Febvre, professor of history in the University of Strasbourg, in a work published in France in 1922, translated into English and published in 1924 as *A Geographical Introduction to History*. (A fourth impression of this book appeared in 1966.) Febvre tells us that the plan of the book was conceived in 1912–13, and though it should have been published in 1915, it did not appear until 1922. This book is a faithful critical reflection of the Vidalian concept of human geography and is an important work in the history of geography. It is a matter of the greatest significance that the new geography was (and continues to be) subjected to the searching analysis and appreciation of both sociologists and historians.

It should be reiterated that the work of Ratzel on human geography was the springboard for Vidal de la Blache so that both directly and indirectly Ratzel's work probably had a far greater influence on future developments in France than in his own country, where Penck, Hettner and Schlüter had a much stronger direct influence on their successors. There was no such genuine direct evaluation of the works of Ratzel in either Britain or America. All that was known seems to have been derived from the writings of Ellen Semple. Pseudo-philosophical discourses were published from time to time between the wars in Britain, but they made no substantive impact on research trends and were preoccupied by the theme of 'possibilism', which has been surpassed for more than a generation.

We now turn briefly to the genealogy of geography in France. Vidal de la Blache dominated the scene in the University of Paris for nearly forty years. The first half of his career covers the last quarter of the nineteenth century. Although a contemporary of Richthofen, Ratzel, Partsch and Wagner, he was their junior and in this period of apprenticeship he learned much from them. The latter half of his career was in the first quarter of the twentieth century and he was thus virtually contemporary with Penck and Hettner. This period also saw the florescence of his school with the publication of massive doctorate studies by his pupils of particular areas of France. These pupils, having served their apprenticeship under him, continued and developed *la tradition vidalienne* after his death in 1918. Thus Vidal de la Blache was the leader of the first generation though he had distinguished contemporary colleagues in Elisée Reclus, De Margerie, the geologist, Franz Scrader, the map-maker, Lucien Gallois, his earliest student and his successor at the Institute, and in Himly, the historian. The outstanding pupils of Vidal de la Blache, who dominated the professional scene in the second quarter of the twentieth century, make up the second generation in France. They include, in addition to Lucien Gallois, the following—Jean Brunhes, Em. de Martonne (Vidal's son-in-law), Albert Demangeon, Maximilien Sorre, Raoul Blanchard, Jules Sion, Henri Baulig, Camille Vallaux, and René Musset. After World War I geographers were installed in all sixteen Universities and were almost without exception pupils of Vidal. During the inter-war period Raoul Blanchard developed a vigorous school at Grenoble and this became a second pole of attraction and dissemination in the profession of geography. By the outbreak of World War II men with doctorates from Grenoble and Paris were to be found in schools and Universities throughout France. These younger men and their associates belong to the third generation.

The third generation includes men who are now nearing, or are already in, retirement. They include André Cholley and Georges Chabot (both former directors of the Institute of Geography in Paris), Pierre George, Roger Dion, and Pierre Deffontaines. Among these must also be listed Ph. Arbos (Clermont Ferrand), A. Allix (Lyon), D. Faucher (Toulouse), and Ch. Robequain (Paris), all of whom completed their doctorate work under Blanchard at Grenoble. Probably in this third generation should also be placed Jean Gottmann, since he was Demangeon's last assistant in the four years immediately preceding the master's death, although Gottman's impact is strong today among his colleagues of the fourth generation.

The fourth generation are the younger geographers who are now well established, but are young enough to have twenty or thirty years of leadership ahead of them. The chief of these are Jean Gottmann, Jaqueline Beaujeu-Garnier, Et. Juillard, Ph. Pinchemel, Max. Derruau, P. Birot and Jean Labasse. There are others, but these names are given to make clear the sequence in the genealogy and the way in which it ties in with Germany, where modern geography had an earlier start by one generation, and where, also there have been many more professional geographers gravitating around several schools in different parts of the country, in distinction to the closely knit field of thought and training under one man that developed in France.

It should be noted, in general conclusion, that over the past twenty years institutes in the provincial Universities have developed a much larger measure of independence of Paris. The preeminence of Grenoble has waned with the retirement of Blanchard but it

Introduction

remains an active seat of research with its own examination system and publications. The same is true of other Universities, particularly for example, Clermont Ferrand, Lyon, Lille, Strasbourg, Bordeaux and Toulouse. There have also been significant developments in research, but these will be reserved for the forthcoming chapters.

Before turning to the work of Vidal de la Blache and his successors, I shall examine the contributions of his contemporaries, giving priority of treatment to the sociologist, Frédéric Le Play.

NOTE

A note on the institutions of higher learning in Paris is necessary here. The *Collège de France* was founded by the State in the sixteenth century as an independent institution of advanced learning. Its professors are engaged in research and give no formal lectures. The *École Normale Supérieure* was an independent institution until 1904 when it was attached to the Sorbonne. Its students are lodged and fed in the school and have some tuition there, but they also attend lectures at the Sorbonne. This institution, of course, is for the training of teachers and Vidal spent many years there. The *Sorbonne* is the old name of the Faculties of Letters and Sciences of Paris and is divided into sections. One of these is the Institute of Geography, that occupies a separate building in the Rue St. Jacques which was erected immediately after World War I when De Martonne became its Director. There has been no professor of geography (as a successor to Vidal) at the École Normale since 1904. The professors of geography today at the Collège de France are R. Dion and P. Gourou, and at the Institute there are ten professors. With the Institute there are associated the *École Supérieure de Cartographie Géographique*, *Le Centre d'Études Supérieures de Tourisme*, and *Le Centre de Recherches et Documentation Cartographiques et Géographiques*.

Further references

In addition to the works cited in the text, the most useful summaries in English are as follows:
R. J. Harrison Church, 'The French School of Geography', in G. Taylor (ed), *Geography in the Twentieth Century*, London, 1951, pp. 70–90.

France

A. Perpillou, 'Geography and Geographical Studies during the War and the Occupation', *Geographical Journal*, 107 (1945), pp. 50–7.

J. Gottmann, 'French Geography in Wartime', *Geographical Review*, 36 (1946), pp. 80–91.

D. V. McKay, 'Colonialism in the French Geographical Movement, 1871–81', *Geographical Review*, 33 (1943), pp. 214–32.

16

Frédéric le Play (1806-82)

We have already emphasized that geography in Germany fell to pieces after 1859 for two or three decades. The regional concept as developed in the first half of the century, as a goal of scientific inquiry, fell into the background in the second half of the century, and did not reemerge as the fundamental task of geography until around 1900. For fifty years, the study of lands and peoples became almost entirely analytical and students of man, notably historians, were primarily interested in the influences of what they called 'geography'—meaning the physical earth—on human character and human history. The study of human societies in their environmental settings rested for some time in the hands of ethnographers. Especially important was the work of Frédéric Le Play and his followers in France, and of distinguished ethnographers both in Germany and in Britain. We shall deal first with the Le Play school. Frédéric Le Play is the only non-geographer who appears in this book as one of its makers. The reason for this is that his thought and methods, for good or ill, have had a profound effect on the development of geography in France and Britain, though his influence was negligible in Germany and the United States. There was a direct impact in Britain through Patrick Geddes and H. J. Herbertson and their contemporary geographers. Lewis Mumford, the foremost follower of Geddes, has developed his ideas in America, as have certain American sociologists, notably P. Sorokin and C. C. Zimmerman, but the impact on geography generally in the States has been negligible. Without evaluation of the work of Le Play, however, an understanding of the development of geographic thought in Britain in particular, would be very inadequate. It is for this reason that we devote the following section to the life and work of Le Play.

Pierre Frédéric Guillaume Le Play,[1] born at Honfleur, Normandy,

[1] The bibliography of the Le Play school is extensive, but apart from the principal separate publications mentioned in the text, the main contributions are in *La Science Sociale*. Particular mention may be made of *La Science Sociale*

in 1806, was educated at the Collège du Havre, and graduated as a mining engineer at the École des Mines in Paris in 1832, where, in 1840, he became Professor of Metallurgy. Le Play was deeply affected by the social upheaval of 1830, which occurred at a time when he was suffering from a serious illness. As a result, he came to believe that it was essential to undertake a scientific examination of social phenomena in order to suggest measures of social reform. To this task he devoted the leisure of his long life. In a professional capacity, and also during his vacations, Le Play travelled through Europe and devoted himself assiduously to the first-hand investigation of social life and organization. The fruits of these labours, spread over twenty years, at length appeared in 1855 in his classic study *Les Ouvriers Européens*. On the basis of these investigations and the conclusions drawn from them regarding the structure and well-being of society, he propounded his schemes of social reform. These are mainly contained in *La réforme sociale en France* (1864) and *L'organisation du travail* (1870).

Le Play also founded two international societies, the names of which indicate their objectives—the International Society for Practical Studies in Social Economy (1856) and the Union of Social Peace (1872). In 1881, the year before he died, he commenced publication of a periodical, *La Réforme Sociale*.

Frédéric Le Play is considered by the distinguished American sociologist, P. Sorokin to be among the 'few names of the most prominent masters of social science'.[1] His work had a profound influence on geographers in France and Britain. Le Play's method of investigation was elaborated and given wider application in France by a group of disciples, the most well-known of whom were Henri de Tourville (1843–1903) and Edmond Demolins (1852–1907). They founded the *Société Internationale de Science Sociale*, and its review *La Science Sociale*.

Le Play felt the lack of both a unit of measurement and a scientific inductive method for the systematic examination of social phenomena. These two problems he solved by taking the family budget as the quantitative expression of family life to be employed as a

[1] P. Sorokin, *Contemporary Sociological Theories*, New York, 1928.

---

d'après Le Play et De Tourville, by F. Champault in *La Science Sociale*, 1913, and *Emm. De Curzon, d'après les principes de Le Play*, 2 vols, Paris, 1897. The following is an important recent publication: *Recueil d'Études sociales à la mémoire de Frédéric Le Play: Centenaire de la Société d'Economie et de Science Sociale fondée par F. Le Play en* 1856, published by Centre Nationale de la Recherche Scientifique, Paris, 1956. This is a collection of papers on the work of Le Play, but also includes substantive studies on current problems.

basis for the analysis of social facts. Furthermore, he recognized that family life and organization are dependent on methods of obtaining subsistence, that is work; while the character of the latter is largely determined by the nature of the environment, that is place. Thus we have the famous formula which is the essence of Le Play's monographic social studies—place, work, family. The family is the unit of its corresponding society, and in order fully to understand the family its interrelations with wider social groupings and institutions must be considered.

In *Les Ouvriers Européens*, each of the monographs on fifty-seven typical families, falls into three sections: (1) Preliminary observations under the headings of (a) environment, (b) means and (c) mode of existence (standard of living), and (d) history, organization, religion, and customs. (2) The budget, in which are entered and compared, the gross receipts and expenditure of the family, 'the balance expressing the standard of living of the family, and above all the moral level which it has attained'. The receipts comprise: (a) fixed and (b) moveable property, (c) wages, and (d) profits from industries and production. On the other side of the ledger expenditures include: (a) food, (b) housing, (c) clothing, (d) moral needs and recreation, and (e) industries, debts, taxes, and insurance. (3) The various elements of social constitution include the wider social institutions which affect the family life. This section is the least systematic part of the monographs.

Le Play and his followers distinguished family types by the laws of inheritance and the character of the education given to the children. Thus, Le Play recognized three main types, each of which was associated in origin with certain physical environmental conditions. The patriarchal family was characteristic of the steppes of Eurasia. The unstable family associated with the forest showed a lack of tradition and of respect for family authority, and a dependence on the state. An example they gave was France. The particularist family (*la famille—souche*) is localized in the fiords of Scandinavia. The individual members of this family type were trained to initiative and on reaching maturity become independent and responsible.

De Tourville, Demolins, and others elaborated Le Play's scheme of social analysis to form *La nomenclature de la science sociale*, an 'instrument of social dissection', with the family as the starting point, then broadening to consider society and its role in the world as the final stage of investigation. The 'nomenclature of social facts' falls into twenty-five divisions (with many subheads). These are grouped under the headings of place, economic factors (labour), and family organization. There follows a treatment of the wider social environment of which the family forms the basic unit, and concludes with the role and future of the society in the world.

Many monographs of social groups were published on this model in *La Science Sociale*, with the 'nomenclature' as their framework. Ed. Demolins, in particular, submitted such a treatment of the 'simpler' societies of the world. Here, the keynote of the almost proforma treatment is the 'determinism' of social structure by geographical conditions (meaning place), a point of view which is now considered to have been greatly overemphasized. The family types, and the development, merits, and defects of their respective societies were thoroughly investigated. The particularist family was considered to be ideal for social progress, and great efforts were made to spread its system and ideals, particularly in France.

Sorokin appraised the nomenclature, listed in the above table, as 'a great contribution to the method of social science'.[1]

Le Play's broad portrayal of the socio-geographic structure of human societies, based on his detailed budgets of type-families, is contained in Volume 1 of the six volumes on *Les ouvriers européens*, which was first published in 1879. It is an important step in the geographical study of human societies and Le Play's argument is briefly outlined.[2]

Le Play postulated that the development of the European peoples took place in three very different 'geographical environments', namely, the steppe, the maritime shores, and the forested lands. The Asiatic steppe was the home of stable nomadic families under the control of patriarchs. On the maritime shores of Europe with their fishing resources, the boat and the habitation were the patrimony of the family which was made up of parents, all unmarried children, and the eldest married son with his family. This is the home of the *famille-souche*. Forested land, covering great areas of much variety, with grass openings, heath, and varied soils, was the birth-place of the unstable family (*la famille instable*), that had also developed in the urban environments of Europe and had spread to America.

Le Play saw the industrial revolution as creating two kinds of

[1] Ibid.

[2] An abridged edition of Volume 1 of *Les Ouvriers Européens*, dated 1879, is available in C. C. Zimmerman and M. E. Frampton, *Family and Society: A Study of the Sociology of Reconstruction*, New York, 1935, pp. 361–595. This translation 'is reduced from a large volume without destroying a single idea'. It is a disquieting thought that this is all thousands of students of sociology will know of geography, an important but completely outmoded approach that was prepared nearly one hundred years ago.

society, one based on the traditional patterns, the other affected by the impact of the new ideas developing in Britain, France, the Low Countries, and Germany. He believed a prosperous, happy society to be dependent on belief in God, practice of religion and acceptance of the *decalogue* (the ten commandments), the authority of the parents, a sovereign government, private property, and a system of mutual help.

Le Play sought to evaluate society on the basis of eight broad functional types. These, recognized on the basis of their means of subsistence (*moyens de subsistence*), are described as follows: savages, shepherds, coastal fishermen, woodsmen, miners and metal workers, agriculturalists, manufacturers, traders and members of the liberal professions.

The importance of this presentation is that it attempts for the first time to classify societies, on the basis of sample family studies, into a number of types and to localize them in relation to their natural environment. It was also argued that in such societies there is a tendency for the mode of livelihood to be reflected in a variety of common social characteristics and in the *mores* of the family. This idea remained a basic objective of geographic research and, as will be seen later, was given special emphasis in the next generation by Paul Vidal de la Blache as he developed his idea of the *genre de vie*.

A classic illustration of the studies that followed Le Play's lead is Demolin's *Les grandes routes des peuples* published in 1901 and 1903.[1] In it Demolins presents an analysis of the social structures of the world's peoples on the basis of what he calls their 'geographical environment', their resultant type of work, and their resultant type of social organization (based on the family unit). A historical inter-pretation runs through the work. Groups develop or become mar-kedly characterized in a particular geographical (physical) milieu and then move outwards, so that their social systems undergo changes in new and different environments. The environment rigidly controls the structure and development of peoples. This was a most extreme form of geographic (environmental) determinism.

The basic idea of Demolins is expressed in the preface to the first volume as follows:

> The primary and decisive cause of the diversity of peoples and races is the *route which has been followed by the peoples*. It is the route (the environment) which created race and social type. . . . It

[1] E. Demolins, *Les Grandes Routes des Peuples: Essai de Géographie sociale, Comment la route cée le type social*. 2 vols. Paris, 1901 and 1903.

has not been an indifferent matter for a people which route they followed: that of the Great Asiatic Steppes, or of the Tundras of Siberia, or the American Savannas or African Forests. Unconsciously and fatally these routes fashioned either the Tartar-Mongol type, Eskimo-Lapps, the Red-Skin or the Negro. In Europe, the Scandinavian type, the Anglo-Saxon, the French, the German, the Greek, the Italian, and the Spanish are also the result of the routes through which their ancestors passed before arrival at the present habitat. ... Modify one or another of these routes ... and through that you will change the social type and race.

To illustrate this approach we may summarize Demolins' treatment (with all its shortcomings) of the nomad societies of the steppes of the Old World. The omnipresence of grass 'determines a uniform mode of labour and pastoral art'. The horse is the chief animal to which 'the steppes are exclusively well adapted, and it is the horse which adapts the steppes to man'. The horse is used for transport and milk (*koumys*). Food, shelter, hygiene, and recreation are determined by the character of the steppe, and labour is almost entirely by hand. The steppes also put their imprint on the character of property and the nature of the family in nomadic society. Property is held in common except when it is cultivated. The characteristic patriarchal family is subject to the rule of the patriarch, with everything in the common ownership of the family. The family is self-sufficient and its members dependent on it are strikingly lacking in initiative and are conservative. The only form of organization is the caravan which could easily be converted into an army endowed, through the use of the horse, with great mobility (Attila, Genghis Khan, and Tamerlane), but the looseness of the social organization explains why these empires so quickly disintegrated.

These peoples spread over much of the world, and one of the most permanent features of their social system was the patriarchal family. One wave spread to the *tundras*, and there the social structure changed in a new environment to give the Eskimo and the Lapp societies. Another branch spread across the Bering Strait to people North America. In new environments, there emerged hunters of the prairies and of the forests, though the patriarchal family survived. A second route of migration in North America was into the mountainous areas of the Rockies, and a third in the forested lands of the Great Lakes. A fourth and most miserable group were driven south into the forest and mountain fastnesses of South America. Under forest conditions an *unstable family* developed, and the hunter and his wife, for example, were forsaken by their children as soon as they were able to take care of themselves.

The particularist family and its society was given special attention, with respect to its origin and spread. This was the special field of Henri de Tourville. The type developed in the fjords of Scandinavia. The fisherman with a tiny patch of cultivated land could not live with a grown family on a restricted food base. The married children left the family and took to fishing and cultivation on another site, for they were dependent entirely on self-help, hence, the development of self-reliance and initiative. This led also to the private ownership of land and government by election and contractual associations on a voluntary basis. This society spread from its Scandinavian base in two directions. From it there developed the Germanic peoples and the growth of feudalism, and a second stream moved across to Britain, and thence the basic social institutions moved centuries later into areas of overseas colonization. This idea later became associated with certain distinctive traits of the so-called Nordic racial types and led to the cult of Nordic racial superiority in the twenties (notably the writings of Madison Grant and Lothrop Stoddara in the States) and under the Nazi regime.

Sorokin summed up the Le Play school as follows:

As yet there has been no sociological school which shows the functions, the classification, and the social importance of the family as clearly as the Le Play school, with the exception of Confucius and the Confucianist school in China. . . . First, the system and the program of the school do not cover the whole field of social phenomena and social problems; second, the school underestimates the factor of heredity and race and overestimates the factor of geographical environment; and third, many problems, analysed by the school, among them the origin of the types of family and the correlation of the types with the social system and historical destinies of a corresponding society, are not quite sufficiently explained. Finally, the applied program is ineffective.[1]

These criticisms, however, should not detract from the great value of the contribution of Le Play and his successors, notably Henri de Tourville, to the interpretation of human societies. The views of Le Play and his followers had a profound influence on the study of human societies in Britain at the beginning of this century. Yet, this school of thought, that was appreciated and used by American sociologists, such as Zimmerman and Sorokin between the wars, has not had the slightest impact on the development of

[1] P. Sorokin, op. cit. p. 89–97.

human geography in America. I take the liberty, for this reason, of commenting briefly on its impact in Britain.

Patrick Geddes,[1] who had a tremendous influence on students of society a generation ago, was greatly influenced by the ideas of Le Play. Geddes in turn influenced the founders of geography in Britain, with whom he closely associated—A. J. Herbertson (who began as an assistant to Geddes at Dundee) and H. J. Mackinder; the early social scientists, such as the Webbs, Graham Wallas and Charles Booth (as for example in his use of the family budget in his survey of London's life and labour); and the greatest of Britain's city planners, Raymond Unwin and Patrick Abercrombie. Many of the ideas formulated by Geddes were adopted and further developed by geographers in Britain. This applies, for example, to his idea of a 'conurbation', that was more precisely defined by C. B. Fawcett, the present writer's teacher and chief for twenty years. The 'valley section'—a river valley from mountain source to coastal estuary—diagrammatically embraced the fundamental occupations of human societies, and for many years was the emblem of the Le Play Society, that grew out of the Sociological Society, with which Victor Branford and Alexander Farquaharson were associated with Geddes. The idea of field study—observation and recording in the field—was in Geddes' teaching basic to the diagnosis and treatment of the maladjustments of social groups in both country and town. Here again one has no difficulty in detecting the adoption of concepts of Le Play. From this idea there emerged what Geddes called 'regional survey', which embraced the trilogy of Place, Work, and Folk, alternatively described as Geography, Anthropology, and Economics, or again as Environment, Function, and Organism. This notion and its procedures served as a medium of education notably in the teaching of geography, as well as a framework for the town and regional planning schemes submitted, mainly under the guidance of Abercrombie, between the wars. This idea was one of the main stimulants to field studies in both training and research in the natural and social sciences, and, unlike the trends in America, is being used in a continually increasing measure in the postwar years as an essential educational medium.

In all these trends, initiated by the work of Patrick Geddes, one can detect clearly the further development of the essential ideas of Le Play. They may have had deterrent effects on the development of

---

[1] See in particular J. Mogey, 'La Science Sociale in England', in *Recueil op. cit.* (fn. 1.), pp. 57–64. Philip Boardman, *Patrick Geddes: Maker of the Future*, Univ. of North Carolina Press, 1944; D. Herbertson, *Frédéric Le Play*, Ledbury, 1950.

geography (as for example in the continuance of 'regional survey', and the continued use of 'the geographical background' as the physical elements only of the area in question), but they certainly mark a period of profound importance in the development of knowledge in Britain, and of geography in particular.

In the United States, the Le Play method of selecting type families in relatively undisturbed rural societies was practised by the sociologist C. C. Zimmerman, for, as already noted, one half of his book on the *Family* follows exactly the Le Play system of family enquiry in backward areas of the States. Lacking presumably appropriate data for Europe, he translated most of the first volume of his *Les Ouvriers Européens*, directly from the master's writings, to illustrate the variations of socio-economic structure in that continent. This book was published in 1935.

There is a second thread in the United States, that connects directly with Geddes and through him with Le Play. The American social philosopher, Lewis Mumford, has professed throughout his life to be a follower of Patrick Geddes and in Mumford's writings one frequently finds ideas and terms that are derived from these sources, and often with acknowledgement. Such terms are 'the valley section', the eight pieces on the chessboard of nature (Le Play's social types), the paleotechnic and neotechnic phases in the historical growth of human societies (to which Mumford added the eotechnic phase, in which technics depend on human and animal muscle, wind and water). These ideas and their development are to be found particularly in *The Culture of Cities* (1938), *Technics and Civilisation* (1934), and *The City in History* (1963). It is a matter of the highest significance that geography in America, particularly in its interpretation of societies, shows, throughout its modern growth, absolutely no recognition or influence of the teaching of Le Play, Geddes or Mumford, great scholars in each of three successive generations. The current works of Mumford provide a clear philosophical and practical direction to the regional concept, which it is the fashion to frown upon, and a corrective to environmental determinism, so much inveighed against, in the light of the technics of Man through history and in their variations in space. This, to the present writer, is one of the enigmas in the development of geography in America.

The influence of Le Play, intermingled with the later ideas of Vidal de la Blache (whose work will be later discussed) is strongly evident, as I have stated, in the early development of geography in Britain. I have good reason to recall this, for I was introduced in school in my early teens, during World War I, to two remarkable works written by two founders of geography in Britain, both of whom were steeped in the European tradition. I refer to A. J. Herbertson and Marion Newbigin. These books were short and simply written, for they were intended for use in schools, but they are both scholarly contributions to the advancement of geography.

A. J. Herbertson and his wife, Dorothy, wrote the first work on human geography in Britain, with the title *Man and his Work*, published in 1902. The authors make specific mention of their indebtedness to Patrick Geddes, and through him to Le Play, 'ably represented at present time by Mr. Edmond Demolins, Director of *La Science Sociale*'. The teaching of these men is abundantly evident throughout the book. It can confidently be asserted (and Fawcett confirmed this to me personally many times) that the men who trained under Herbertson at Oxford received the same imprint. We read in this book phrases such as the following: 'The world is the home of man. All that we learn of the physical features of the Earth, its climates, and animals is of practical importance, because these things have helped to make the human race what it is—here adventurous and progressive, there indolent and backward.'

Life is first outlined in the framework of the great vegetation zones—frozen desert, temperate forests, steppes, hot deserts, equatorial forest, mountains, plains, and coasts. Then follows a section on the influence of occupation, in the sense that 'the same occupations tend to produce societies of the same general type'. He then examines hunting societies, fishing societies, pastoral societies, agricultural 'races'; and the influence of occupations on dwellings and clothing and food. Agriculture is examined in terms of crops and their distribution, and the stages in the development of agriculture. There follow chapters on the 'rise of the arts', the 'development of manufactures', 'trade and transport', trade routes and towns, the distribution and movement of population (densities of different occupational groups, expansion and migrations), and the book concludes with a brief survey of the 'races of men'.

Marion Newbigin's little book on *Man and His Conquest of Nature*, was published in 1912. She acknowledges her debt to the writings of the French geographers, and mentions by name Brunhes, De Martonne, Vidal de la Blache, and Demangeon; their works will be discussed in later chapters. The focus of the book is on 'the relations between groups of men and their surroundings', which, at this date, was a rare presentation and as she claimed had barely reached 'the ordinary text book' used in schools. The central geographical question, she states, is: 'Why is it easier for men to make their living at some places than at others?'.

Frédéric le Play

A listing of the contents by chapters reads as follows: The Battle-field; Examples of advanced communities: (i) self-sufficing groups, and (ii) dependent groups; Primitive conditions: the collecting stage; Man's helpers: (i) domesticated animals, and (ii) mechanical aids and means of transport; Cultivated plants: primary products, and those yielding luxuries or secondary products; The useful minerals and the Industrial Revolution; Some communities outside the Coal Zone; The harvest of the Sea; and Regions where Man has thriven.

This book, in effect, is a presentation of Vidal's idea of the mode of life (*genre de vie*) and is a direct descendent of Le Play's classification and treatment of the world's societies. Although intended as a school book, the presentation embodies an approach that was much more mature than the theoretical statements of her contemporaries. The idea of modes of livelihood in relation to the environmental setting was being developed at this time by Roxby and Fleure, the professional leaders of geography in Britain at this time, and both of them deeply steeped in the tradition of Vidal de la Blache, Geddes and Le Play. One is mindful in particular of Fleure's conception of world areas in regard to the problems and difficulties posed by the conditions of the physical environment. Thus, he wrote in 1916 of regions of debilitation, regions of incre-ment, regions of effort, regions of difficulty, etc. These areas, though crudely expressed (the purpose was to advance the concept) have been defined in various sophisticated ways since this essay was first published and the student of levels of economic health, levels of area development, in terms of criteria and areal definition is probably not even aware of the antecedents of this mode of geographic ap-proach and its contribution to the understanding of world problems.

Newbigin based her book, as did her French colleagues, on the basic needs of food, clothing and shelter. In the 'spread of man over the earth', he has come into conflict with Nature and with other human groups and their ideas. She tells us in the concluding paragraphs that:

It is no part of the work of geography to lay down rules as to the best solutions of particular problems of land utilisation, but it is her work to set forth as clearly as may be the conditions which reign at different parts of the earth's surface, for a consideration of these furnishes the raw material upon which all political and social judgments must be based.

These are wise words indeed and were about a generation ahead of their time.

17

Vidal de la Blache (1845-1918)

The growth of geography in France, unlike that in Germany with its several distinct schools of thought, has been shaped by the work of one man, namely, Vidal de la Blache. For a generation his school was dominant and, though the goals of the founder are today apparently somewhat outdistanced, French colleagues now refer to *la tradition vidalienne*. This chapter is devoted, therefore, to the works of Vidal de la Blache and to one of his early pupils, Jean Brunhes who clarified the field of study his teacher so clearly established.

Paul Vidal de la Blache (1845–1918) graduated in history and geography in 1865 at the *École Normale Supérieure* in Paris. After spending some time teaching at the French School in Athens, he returned to France in 1872 to obtain a doctorate. He joined the faculty of the University of Nancy and spent five years there, working on the area, from which emerged his study of *La France de l'Est* (published in 1917, towards the end of his life). At this time he also visited Germany where he met Richthofen. In 1877 he transferred to the *École Normale Supérieure* in Paris. He moved to the Sorbonne in 1898 and remained there until his death in 1918. For over thirty years he pursued his studies in geography and inspired many students who subsequently became professors in universities in France. W. L. G. Joerg wrote in 1922 'Nearly all the occupants of chairs in geography in France are pupils or pupils of pupils of the late Vidal de la Blache. In no other country, it may be said, has the development of geography centred about one man as in France.'[1]

Vidal de la Blache wrote outstanding contributions to the field of geography. These include basic articles on the 'geographical conditions of social facts' (1902), the 'distinctive characteristics of geography' (1913), and 'modes of life' (*genres de vie*) in human geography (1911), all published in the journal he founded, the *Annales de Géographie*. We are fortunate in having some of these and other

[1] W. L. G. Joerg, 'Recent Geographical Work in Europe', *Geog. Rev.*, 12 (1922), pp. 431–84.

Vidal de la Blache

materials collected by Em. de Martonne in one volume with the title *Principles of Human Geography*, published in 1921. This book deals with the world-wide distribution of population and settlements; types and distributions of civilization (a term which our contemporaries would prefer to call 'cultures'); and the distribution and development of various forms of transportation. Rural settlements, their groupings and the forms of the buildings, 'modify the landscape profoundly, and form on that account one of the essential elements of human geography'. The same idea applies to fields, routes, cities, woods and other elements of the landscape.

Vidal de la Blache insisted on the study of the physical earth—Nature—which he referred to as the geographical environment—and of Man. He considered, however, that the peculiar obligation of the geographer lies in the correlation of physical and human conditions in their spatial interrelations. The physical (natural) environment provides a range of possibilities which Man turns to his use according to his needs, wishes and capacities, in creating his habitat. Hence, as opposed to the 'environmental determinism' that was prevalent at the time as a Darwinian heritage, he established a conceptual framework of 'possibilism' that was fully developed by a critical historian, Lucien Febvre.[1] Through his occupance and imprint on the land, Man creates distinctive countries, be they states or minor unit areas (*pays*). Man establishes relations with the environment not as an individual, but through the heritage and objectives of the group to which he belongs. It will be apparent that this whole approach clearly demands historical interpretation in order to evaluate the changing modes and objectives of human adjustment. It also demands a thorough appraisal of the natural environment, both as an end in itself (physical geography proper) as well as in its relevance to human occupance.

Other works by Vidal de la Blache may be noted. *Tableau de la Géographie de la France* (1903) is regarded as one of the classics of geographical literature. It shows a preoccupation with the physical features of the earth, man's imprint upon the landscape, and the organization of the country into distinctive units. The preparation of an atlas of geography and history (1894) demanded several years of work early in his career. The volume on *La France de l'Est* was his last work (1917) although based upon his earliest researches. He founded the *Annales* in 1891 in collaboration with M. Dubois, the colonial geographer, and scarcely a number passed without his having a contribution in it. The *Géographie Universelle* was conceived

[1] Lucien Febvre, *A Geographical Introduction to History*, London, 1924, an English translation.

and organized by him, and this third production of a *Géographie Universelle*, after the first and second efforts by Malte-Brun (1810–29) and Élisée Reclus (1876–94), stands as a monumental exposition of the lands and peoples of the world. Finally, Vidal de la Blache attracted a number of ardent pupils who produced, under his inspiration, 'regional monographs', many of which were published before World War I. This tradition has continued in the training of the professional geographer and detailed studies of lands in France as well as overseas (particularly in former 'colonial' territories) are a formidable contribution to modern knowledge that owes its beginnings and continued inspiration to the founder. Special attention will be given to the monographs in a later section.

Vidal de la Blache sought to establish geography as a distinct discipline.[1] Its field of study is the 'ensemble of phenomena that occurs in the zone of contact of the solid, liquid and gaseous masses that make up the planet'. (Recent German scholars refer to this as the 'geosphere'.) These phenomena are studied in relation to 'place, localization and distribution'. Thus, the field of study has a double aspect—Nature and Man. To explain their expression on the earth's surface one needs to borrow from both the natural sciences and humanities, especially from geology and history. As for Nature, the surface of the lands must be interpreted in the light of the geological past, never losing sight of 'the outlines of the countryside, the forms of things'. The focus of study is *la physionomie des paysages*, and this means the real, visible, landscape or habitat, that is the product of both Nature and Man.

Vidal de la Blache considered the physical milieu as the 'natural' or 'geographic' environment of human occupance. This was the current viewpoint and it is still widely held, though it is a misconception, or certainly a misnomer. Indeed, he wrote in 1902 of 'the geographical conditions of social facts', and one certainly gets the impression that the latter should be examined only in so far as they enter into connection with the 'geographical' (i.e. natural) environment.

Social conditions find their expression in distinctive modes of life (*genres de vie*), which in simpler societies, show a close adjustment to the natural environment. Vidal de la Blache refined the views of Le Play and, in English, these ideas find their parallel in Herbertson's *Man and his Work*, published in 1902. The approach is evident in the attention given to the man-made forms of the landscape—roads, fields, farms, villages and towns; to the distribution and density and

[1] A. Demangeon, 'Vidal de la Blache', *Revue Universitaire*, June (1918), pp. 4–15.

movements of human populations; to the characteristics and distribution of types of culture (as in Cvijic's study of zones of civilization in the Balkans); and to the distribution and environmental relations of food, clothing and material equipment and shelter (dwellings).

An article in the *Annales* in 1913 outlines the 'distinctive characteristics of geography'. He claims that the study has its goal in the characteristics and groupings of the phenomena of the landscape as the expression of man's presence and works in it. It is here that, almost word for word, he repeats the famous dictum of Ritter. While insisting on the need for both geology and history in interpreting the distinctive areal groupings of the earth's surface, he continues:

> That which geography, in exchange for the help it receives from other sciences, can offer to the common treasure, is the aptitude not to break up what Nature brings together, to understand the correspondence and the correlation of facts, be they in the terrestrial milieu which envelopes them all, or in the regional environments in which they are localized.

He lists the distinctive concepts of the discipline as follows:

1. Geography recognizes the unity of terrestrial phenomena. It, therefore, pursues a world view and consideration of causes and generic types of groupings of terrestrial phenomena.
2. Terrestrial phenomena are localized in varied combinations and modifications of different sets of phenomena.
3. Geography seeks to describe, localize and explain these co-variations.
4. Geography seeks to measure the influence of environment, especially climate and vegetation, on man.
5. Geography, therefore, seeks for scientific methods of defining and classifying terrestrial phenomena.
6. Geography seeks to measure and localize the great part that Man plays in modifying the face of the earth.

The same objectives were subsequently pursued by a leading pupil, Albert Demangeon (pp. 231–34). In *Problèmes de Géographie Humaine* (1954) (papers published posthumously) he lists four major fields of investigation for human geography. These are: first, the study of resources, natural and human, as a further development of the concept of *genre de vie* of Vidal de la Blache; second, the evolution and localization of different types of civilization; third, the distribution and movements of population; and fourth, the classification, distribution and development of the facts of the humanized landscape—field, house, village, route, town and state.

JEAN BRUNHES

Jean Brunhes (1869–1930), was an outstanding scholar in France and a maker of geography, his name being as well known as that of Vidal de la Blache through the English translation of his work on human geography. Like his German contemporary, Otto Schlüter, he sought to give the geography of Man a clear disciplinary basis. He studied under Vidal de la Blache, whom he regarded as *le maître*. Moreover, it was the latter who commended Brunhes and his work on human geography to the Academy of Moral and Political Sciences in 1911.

Jean Brunhes, born in Toulouse, studied at the *École Normale Supérieure* in Paris, like so many of the French geographers, with the intention of becoming a teacher. He worked under Vidal de la Blache and took the *agregé* in history and geography in 1892. With the aid of scholarships, he continued his studies in Paris until 1896, attending courses in the *École des Mines*, *École des Ponts et Chaussés*, and *Institut Agronomique*. He was called as professor of geography to the University of Fribourg in Switzerland in 1896, and, after 1908, he gave a course on human geography at the University of Lausanne. In 1894 he spent some time in Spain, and out of this experience emerged his thesis on *Irrigation: ses conditions géographiques, ses modes et son organization dans la peninsule ibèrique et dans l'Afrique du Nord*. This was published in 1902. Brunhes, however, had other interests. In 1891 he translated into French, *La Constitution d'Athènes*, the treatise of Aristotle that was discovered on papyrus a few years earlier in Egypt. In 1898 he received an award from the *Académie française* for a memoire on *Michelet*, and, in 1901, he wrote, together with his wife, a book on *Ruskin and the Bible*.

Brunhes' main work appeared in 1910 with the title *Géographie Humaine: essai de classification positive*.[1] On the strength of this work he entered the *Collège de France*, where, from 1912, he occupied a research chair of human geography. He was asked by G. Hanotaux, the distinguished historian, to write the first two volumes

[1] *La Géographie Humaine*, 1910, Paris, 3 volumes, 4th ed. 1934 translated by I. Bowman, R. E. Dodge and T. C. LeCompte, as *Human Geography*, Chicago, 1920, 1 volume. A French abridged edition, prepared by his daughter and Pierre Deffontaines, appeared in 1942 (2nd. ed., 1947).

of *Histoire de la Nation Française*. This work carried the title of *Géographie Humaine de la France*,[1] and was written in the early twenties with the collaboration of P. Girardin and P. Deffontaines. He also wrote, in collaboration with C. Vallaux, *La Géographie de l'histoire: Géographie de la Paix et de la Guerre sur terre et sur mer* (1921). On the merits of all these works he was elected to the *Academie des Sciences* in 1927. In 1928 he translated, and brought up-to-date (with the assistance of his daughter), the work on political geography of the American geographer, I. Bowman, with the title *Le Monde Nouveau*. He died in 1930. The translation of Bowman's work into French, and of Brunhes' work on human geography into English is one more indication of the close *rapport*, both professional and personal, between the leaders of American and French geography in this period.

Brunhes' outlook was inspired by the teaching of Vidal de la Blache, but it inevitably shifted the emphasis from the focal study of 'geographical conditions' and 'social groupings' to the study of the surface features of the earth. Brunhes elaborated a system of 'human geography' with its raw material in 'the visible and tangible facts' of human activity on the earth's surface. These are called the essential facts of human geography. Similarly, if one adds (as one must) the physical facts of the landscape—surface configuration, vegetation, soils in their surface expression as forms of terrain (and the term, *les formes du terrain*, is none other than that used by De Noe and De Margerie in 1887), then we have the essential facts of geography.

The essential facts of human geography are divided into three pairs:
1. facts of unproductive occupation of the soil—houses and roads.
2. facts of vegetable and animal conquest.
3. facts of destructive economy in vegetable, animal and mineral.

These are examined in particular type areas or 'regional studies' to show the interdependence of these phenomena in contrasted physical and social environments.

According to Brunhes, study of these three sets of observable phenomena does not set the limit to geography, a fact that is not generally appreciated. Beyond the essential facts, but in association with them, is an area of wider study. This is described as 'the geography of history', and it embraces five aspects: the distribution of population (static conditions and migrations); economic geography

---

[1] Vol. 1, *Géographie générale et Géographie régionale*, 1920 and Vol. 2, *Géographie Politique et Géographie de Travail*, 1926. The second volume was written in collaboration with P. Deffontaines.

or the geography of work (production, transport, and exchange); the geography of political societies (territory, routes and frontiers, groups of states); geography of civilization or social geography (nationalities, races, languages, religions, intellectual, artistic and technical phenomena and group attitudes (mentality, organization, juridical, social) where these are associated with the 'essential facts'; and 'regional geography (*Länderkunde*, *Corografia*), which is a synthesis of the foregoing'.

The intangible phenomena of society, that are indeed localized in particular areas and thus constitute geographic groupings, are only considered in so far as they are necessary to understand the processes whereby the landscape elements and their areal groupings are formed. These are logical fields of geographic investigation—the distinctive areas of human life and organization—but they lie beyond the field of the essential facts. The latter, argues Brunhes, are the peculiar 'autonomous' field of geography. The former, the ancillary and peripheral areas, may well also be cultivated by other scholars. This was clearly a question of emphasis and Brunhes was equally interested in the peripheral fields. The same stand was taken by C. Vallaux in 1925 in *Les Sciences Géographiques*.

THE REGIONAL MONOGRAPH

Vidal de la Blache insisted that geographical research, and certainly the training of a geographer, should concentrate on the study of particular areas, small and accessible enough for thorough study in the field, among the people, and in the stacks of the archival library. In this way one can arrive inductively at valid generalizations with respect to the lands and peoples of the earth. This is precisely the method and purpose of Ritter's *Erdkunde*.

> If this danger (of generalization) is to be avoided, one must have recourse to antidotes. I can advise nothing better than the preparation of analytical studies or monographs in which the relation between geographical conditions (*sic*) and social facts are viewed at close quarters, in a well chosen and small field.[1]

It is necessary to point out the elementary fact that Vidal de la Blache did *not* mean by 'region' a clearly bounded area in space as a frame of areal description. He made reference to the search for the areal interdependence of terrestrial phenomena within an area that is selected on a broad basis of similarity or uniqueness as a convenient areal unit for operation. The maps in the early monographs are

[1] *Annales de Géographie*, 11 (1902), p. 23.

execrable and never do their writers indulge in defining divisions as a framework *ab initio*. The problem is to discover such areal associations and seek their causes. This is widely misinterpreted among English-speaking geographers.

A. Demangeon, in an appreciation of Vidal de la Blache, wrote as follows in 1918:

> Every region (*sic*) has its unique character to which contribute the features of the soil, atmosphere, plants and man. The aim of all (geographical) research consists in the analysis of these features. The aim of description is to synthesise these and to show the interlocking of all the phenomena which comprise regional types.[1]

The first studies of this new kind, inspired and supervised by Vidal de la Blache, were submitted by Emm. de Martonne on Wallachia in 1902, A. Demangeon on Picardy in 1905, and R. Blanchard on Flanders in 1906. Others to appear in quick succession were: C. Vallaux, *La Basse Bretagne* (1906); J. Sion, *Les Paysans de la Normandie Orientale* (1908); J. Levainville, *Le Morvan* (1909); A. Vacher, *Le Berry* (1908); Passerat, *Les Plaines de Le Poitou* (1909); M. Sorre, *Les Pyréneés Méditerranéennes* (1913); R. Musset, *Le Bas Baine* (1917).

The tradition was vigorously pursued in the inter-war years after the master's death, with, however, some important changes. French geography, writes R. Musset in 1938,[2] has been especially dedicated to 'regional descriptive geography' in *la tradition vidalienne*. Such works, writes Musset, stand as monuments of scholarship in the literature of France and this is no exaggeration. 'Perhaps no other science reveals a greater significance of description'. Yet 'regional geography' is closely allied, on its explanatory side, with the principle of general geography. For, as Vidal de la Blache maintained, exhaustive studies of particular areas must form the basis of sound generalizations. Regional and general geography are interdependent.

The so-called regional treatment is a work of science and artistry in one, says Musset. For a region has a geographical personality. 'The individual character is sought for by recognition of the influences of Nature, the activity of Man, and the results of historical development and the interaction of all these factors.'

Many scholarly regional monographs appeared between the wars. Emphasis depends on the purpose of the worker and the character of the area. Some are limited to physical geography, e.g. P. Birot on the morphology of the eastern Pyrenees, G. Chabot on the plateaux

[1] A. Demangeon, 'Vidal de la Blache', *Revue Universitaire*, 1918.
[2] R. Musset, *Geographische Zeitschrift*, 44 (1938).

of the central Jura, M. Pardé on the regime of the Rhône, A. Perpillou on the Limousin, E. Bénèvent on the climate of the French Alps, and H. Baulig on the geomorphology of the central plateau. Others focus on human geography, as for example Deffontaine's on man and his works in the middle Garonne valley (1932), Th. Lefebvre on modes of life in the Atlantic Pyrenees, P. Marres on the Causses (1935), and M. Perrin on the Industrial Region of Saint-Etienne (1937). Others contain a full study of both physical and human aspects of geography. Examples (with English titles) are: A. Cholley, on *The Pre-alps of Savoy* and their foreland (1925), D. Faucher on *The Plains and Basins of the Middle Rhone* (1927), A. Gibert on *The Gate of Burgundy and Alsace* (1930), A. Allix on *L'Oisans* (1929), J. Blache on *Massifs of the Grande—Chartreuse and Vercors* (1933), R. Dion on *The Loire Valley* (1933), P. George on *The Region of the Bas-Rhone* (1938), and A. Meynier on *Segalas, Levezou and Chataigneraie* (1931). A number of monographs were also written on foreign areas. Examples are Ch. Robequain's *Le Than Hoa* (in the Annam) (1929), P. Gourou, *Peasants of the Tonkin Delta* (1936), R. Thoumin, *Central Syria* (1936), and R. Capot-Rey on the *Saar* (1934). The list is not complete and is limited only to those volumes that I have read.

A change of approach in the regional monograph from Vidal's initial framework is already fully apparent in the twenties. This is revealed in a comparison of Demangeon's monograph on Picardy, published in 1905, and Deffontaines' on the lands of the middle Garonne published in 1932, or Dion's on the middle valley of the Loire, published in 1934. Each of these will be briefly commented upon, though the value and purpose of these works can only be adequately appreciated by the arduous task of reading at least one of them, provided with the appropriate large-scale maps, preferably backed up by first-hand acquaintance with the area of study, and much patience with tedious topographic detail. This has been my long experience, accompanied in some cases in the field with the author himself. It is my firm opinion that the study of one regional monograph, with the appropriate map studies, and ideally followed by work in the field thereafter, should be an essential part of the training of every undergraduate student of geography, either as an individual assignment, or preferably as a co-operative group study.

Demangeon's monograph on *La Picardie*, reflects the conceptual framework of geography that prevailed at the time of Vidal de la Blache. The area is defined in the first pages as a historical association of people who have made a distinct countryside in contrast to the districts around it. Picardy most emphatically is not coterminus with the geological base of chalk. It is a human entity. The presentation then follows the following sequence (by chapter headings): land-forms, soils (chalk, clay—with flints and loams, examined in their relevance to landscape and human occupance); climate; hydrography, again with relevance to human occupance (drainage, level of water table, etc); natural conditions, reclamation, and agricultural occupance of the coastal marshes; agriculture, in terms of the history of localization of forest clearance, size of holdings, crops and livestock; localization of urban crafts (textiles) and the origin, development, and distribution of rural crafts; routes of trade and economic contacts; origin and distribution of communal lands, land ownership and holdings; the disposition, typology, and distribution of settlements—farmsteads (arrangement of individual buildings), villages, townlets, and towns; the distribution of population, density, depopulation, and migrations; and the territorial divisions (*pays*) within Picardy, their locale and limits, and the determinants—historical origins, physical conditions, and association with town centres.

Deffontaines' study (1932) is concerned with the human occupance of the lands of the Middle Garonne. The physical geography is summarized in a few pages simply at the beginning as orientation but it is skilfully woven into the exposition of the changing patterns and forms of human occupance. The changing socio-economic structure through history provides the guidelines. The study begins with the present and works backwards, ending with the settlement of the early middle ages, when the present humanized landscape received its first decisive imprint of human occupance. Here is a refinement and thorough application of the regional concept, in terms of the impact of the changing spatial structure of human societies on the landscape, that was the result of work done in the 1920's, fifty years ago.

Following is the sequence of chapters. 'The human effective' embraces types of habitation (the typology of individual buildings) and clustered settlements, from hamlets to towns, and their territorial framework (communes). The phases of settlement cover (i) depopulation in the nineteenth century; (ii) overpopulation in the eighteenth century; (iii) the recolonization of the *bastides* (small new fortified towns) in the thirteenth and fourteenth centuries; and (iv) the demographic role of the Abbeys in the eleventh and twelfth centuries. Each of these receives one or more chapters. 'The *horizons de travail*' are the dominant modes of occupation. Agricultural occupance begins with the systems prevalent today, and the sequence of treatment goes back to the 'wheat-maize cycle' of the seventeenth and eighteenth centuries, and the systems of cultivation in the middle ages. The industrial horizons of work embrace chapters on their present day character and distribution; and character and locale in the eighteenth century. Commerce is examined in relation to the main transport arteries of the great valleys and their adaptability to transport and the siting of trading sites (river regimes, flood plains, reclamation and utilization); the network of routes by river and road, and modes of life associated with them. A last chapter is devoted to the market centres and their changing role in this transport net.

A historical approach is evident in Dion's study of the Loire Valley published in 1934. It was reviewed by A. Demangeon in the *Annales* (1934) as follows:

We find here physical and human geography in intimate collaboration. The majority of the regional monographs of France have normally contained two parts, physical and human, since they first appeared thirty years ago. But now one finds a tendency, that evidently results from a difference of method, to treat them as a whole (together), but independently. Thus one gets two separate chambers, separated by a narrow *cloison*, that have nothing in common except one roof and builder. But in M. Dion's work the two are constantly interwoven, for all roads lead to the human aspect. This applies, for example, to chapters on climate, the topography of the Val (floor of the river) and the floods of the river itself. He relates continuously the *local* to the *general* (wider) conditions. This is evident in the way in which the two agrarian systems of the north and south of France have intermingled in the area of the Val.

This large work falls into three 'books'. The first deals with the natural characteristics of the Val de Loire—extent, climate, landforms, and the specific characteristics of the river, especially the nature of the river plain and the regime. The second book deals with the history of man's transformation of the river plain and his struggle with the vagaries of the flow of the river. Particular emphasis is therefore given to the construction of the containing embankments (*levées*) through the centuries since the middle ages. The third book deals with the rural life and organization of the Val in terms of the evidences of ancient rural traditions, the agricultural revolution of the nineteenth century, and types of agriculture today, with special reference to the role of the vine in the economy.

The culmination of this regional approach is the *Géographie Universelle* which was published in its entirety in the inter-war period. This great work was initiated by Vidal de la Blache and completed almost entirely by his pupils under the editorship of Lucien

Vidal de la Blache

Gallois. Its 23 volumes conclude with one large volume on the 'physical geography' of France by Emm. de Martonne and two volumes on the 'economic and human geography' of France by A. Demangeon. These works embodied the ideas of Vidal de la Blache and the substance of a generation of workers. Vidal de la Blache wrote the editorial *leitmotif* in these words: *Par dessus le localisme, les rapports généraux entre la terre et l'homme se font jour.*

The volume on physical geography of France (1942) is by Emm. de Martonne. The first half covers surface configuration, which is presented on the basis of *grandes régions*—Paris Basin, Hercynian Massifs, Armorica, Ardennes, Vosges, Massif Central, Alps, Jura, and the Saône-Rhodanian couloir (including Corsica), Basin of Aquitaine, and littoral forms. The second part covers climate, hydrography, and vegetation.

Two large volumes (1946 and 1948) written by A. Demangeon are devoted to 'economic and human geography'. The first volume, after a brief section on the personality of France, is divided into four parts—the place of France in Europe and the world; agricultural economies and rural life; major agricultural regions (West, North, West-central, Massif Central, Basin of Aquitaine, Mediterranean lands, Alps, Pyrenees); circulation (historical development and locale of roads, railroads and waterways).

The second volume falls into five parts. The first is a consideration of the functions of urban settlements in their regional settings, concluding with a study of seasonal towns (*les villes saisonnières*). The second part deals with human activities associated with the sea, and again this is treated regionally (North, Normandy, Brittany, Atlantic and Mediterranean shores). The third part deals with industrial economies that are again dealt with regionally—North, Rouen complex, East, Lyon complex, Centre, Midi, Alps. There follows in the fourth part a detailed study of Paris—formation, growth, population, work, limits and satellites. A fifth part deals with France as a whole in reference to its growth as a politico-geographic entity and its subdivisions. This whole work is treated in terms of France as a whole but interprets the human aspects in their relevant regional contexts. There is no attempt to dragoon all the varied aspects of man-land relationships into a fixed 'regional' framework, as is so commonly done elsewhere.

One finds the phrase *la tradition vidalienne* in use among French geographers in recent years. We shall later revert to it, for there is today a divergence from the impetus given by the strong leadership of the founder. We are here concerned with the trends as they were developed by the pupils of Vidal de la Blache in the first two decades

219

of this century and after his death from 1918 to 1938. Rather than express my own views on this matter, I prefer to quote the words of Jean Gottmann, from a personal letter to me dated 15th January, 1967.

In the official French interpretation, there were two branches of the French school of geography founded by Vidal de la Blache. One was the Paris school, headed after Vidal's death by a team consisting of Emmanuel de Martonne (Vidal's son-in-law and a physical geographer) and Albert Demangeon (human geography). The other branch was the school of Grenoble, headed by Raoul Blanchard. In my opinion the Paris school (to which I belong, of course) was more analytical and systematic, the Grenoble school was more one of 'portraitists', owing to the skill and art of its leader. It is noteworthy that the Grenoble school was very prolific and supplied a large proportion of the faculty in the departments of geography of the provincial French universities in the years 1925 to 1955. Among the students of Blanchard were Allix, Faucher, Arbos, Blache, and Robequain.

The Paris school was in fact more numerous, but a great many of the students of Demangeon broke away from the teaching of geography proper in the French universities, for at least a part of their career. Thus, the Paris school seemed to maintain the tradition chiefly inherited from De Martonne and his immediate students (especially Cholley, who succeeded him in his chair at the Sorbonne), and that tradition was basically rooted in physical geography and descriptive regional geography. A good many of the French geographers now specializing in human and economic matters started with a physical emphasis.

Now let me return for a while to Vidal de la Blache. I had not known him, but I worked at some moments in my career with many of his major students (including Demangeon, my main teacher; Emm. de Martonne, Henri Baulig, Jules Sion, Lucien Gallois, E. F. Gautier, André Siegfried, and Paul Mantou), and of course I met many others among his students. There is no doubt that Vidal started the great tradition of regional studies and also of physical geography in France. But, as you probably know, he was himself a classicist by training, specialized originally on ancient Greece. He went for his doctoral work to the French school of archaeology in Athens. Some ten years ago I was there and found in the records of the school the exact dates of his sojourn and of the topic on which he worked. He discovered there Ptolemy and Strabo, and wrote a rather interesting work on

Ptolemy. That gave him the feeling that this kind of science should also be practiced in the modern universities. But I am convinced that before and beyond the ancient geographers Vidal had been impressed with the basic philosophy of Plato and Aristotle. Whoever has studied Plato's Dialogues and Aristotle's Politics cannot accept environmental determinism as easily as Ratzel and others do. On the other hand, Vidal may have been impressed more than the following generation of classicists were by Aristotle's sophism about 'place' and 'nowhere' in his Physics.

I would say that Vidal's tradition emphasized both some Ptolemaic elements and some Platonic elements. Among his students and followers many picked up chiefly the Ptolemaic line, others were more affected by the Platonic and possibly even Socratic thread in the master's teaching. I think that when you question the cases of Siegfried, Ancel, Robequain and Gourou you picked up people who were particularly affected by the Platonic ideas concerning resources and society.

André Journaux writes as follows in 1966 on the post-war trends in France, that is, in the generation following that of Demangeon, Blanchard and de Martonne.

The majority of researches began in France with the publication of theses on regional geography. After these first investigations, that were a necessary and serious base for a knowledge of the world, above all of France, one has pursued more systematic studies of major phenomena, physical or human.

Since the war, theses embrace much more extended topics, that are relevant to a series of interrelated phenomena in a more limited area. Without attaining always the amplitude of a general synthesis, they contribute in a larger regional framework to a better understanding of the distribution of certain phenomena, physical, biological and human, and the causes of their distribution. It will be for a future generation to take up the elements again of these general studies to compose regional syntheses.[1]

I shall return to this theme on a later page (p. 260) when dealing with post-war trends.

---

[1] A. Journaux, in the *Introduction to Géographie Générale, Encyclopédia de la Pléiade*, Paris, 1966, p. 6.

18

Contemporaries of Vidal de la Blache

ÉLISÉE RECLUS

Élisée Reclus (1830–1905)[1] attended Ritter's lectures in Berlin in 1849–50, but his concepts and writing advanced far beyond what he learned from the master. After being expelled from France as an anarchist in 1851 he travelled and worked in England, in the United States, and in South America, before returning to France in 1856. The two volumes of *La Terre*, which revealed the strong influence of Ritter, were written at this time, though they were not published until 1868–9, and the English translation was done in 1872. Reclus was again condemned to penal servitude as an anarchist at the barricades in 1871, but this sentence was later converted to exile. The nineteen volumes of his *Nouvelle Géographie Universelle* were published between 1876 and 1894. In his last years he lectured in Edinburgh and Brussels where he had an important influence on the development of geography in Belgium. The six volumes of *L'homme et la Terre* appeared in 1905, the year of his death.

The *Nouvelle Géographie Universelle* was translated as the *Earth and Its Inhabitants*, edited in part, by the British anthropologist A. H. Keane, and published in New York between 1882 and 1895. This is one of the great works of geographical scholarship. It is well written and is illustrated by 3,000 maps. I am inclined to say that in it Reclus achieved single handed what Ritter failed to do, in that he produced a description of the earth on a regional basis.

Reclus' descriptive material is lucid, detailed, systematic, and balanced, while his philosophical approach reflects his time. Thus, all the facts of primitive history are explained by the disposition of the 'geographical' theatre in which they have taken place.

We have even a right to assert that the history of the development

[1] M. W. Mikesell, 'Observations on the writings of Élisée Reclus', *Geography*, 34(1959), pp. 221–6.

of mankind was written beforehand in sublime lettering on the plains, valleys and coasts of our continents.[1]

Such extravagant claims are offset, however, by the following: 'Man may modify (his dwelling place) to suit his own purpose; he may overcome nature, as it were, and convert the energies of the earth into domesticated forces.' One must seek the 'gradual changes in the historical importance of the configuration of the land', and in studying space 'we must take account of another element of equal value—time'.[2]

The theme of Man's role in changing the face of the earth, which figures prominently in the writings of Reclus, was adumbrated in various writings before his time. It first found expression in the writings of Count Buffon in the eighteenth century. George Perkins Marsh in the middle of the nineteenth century warned against interference 'with the spontaneous arrangement of the organic and inorganic world'.[3] The same theme was developed at the end of the century by the Harvard geologist Nathaniel Shaler[4] and by the Russian geographer, Alexander Woeikof.[5] It gradually became associated with the deleterious effects of human agency and particularly with the idea of 'destructive exploitation', as later developed by Jean Bruhnes, or *Raubwirtschaft* as it was called by German scholars. Out of this ever increasing and widening impact of waste and depletion there has emerged in this century the concept of conservation of natural resources.[6]

Reclus was much concerned with Man's lack of taste in spoliating the landscape. He was concerned with the destruction of the beauties of Nature, and believed that Man, in developing resources and building his works, should give 'grace and majesty to the scenery'. But, Reclus said, through abuse of his powers 'the barbarian gives to the earth he lives on an aspect of rough brutality', and in extreme cases 'where all grace and poetry have disappeared from the landscape, imagination dies out, the mind is impoverished, and a spirit of routine and servility takes possession of the soul'. Rural life is

[1] *Ocean*, p. 435, quoted by Mikesell *op. cit.*, p. 223.

[2] Élisée Reclus, *The Earth and Its Inhabitants*, ed. E. G. Ravenstein, New York, 1882, Vol. 1, p. 5.

[3] George Perkins Marsh, *Man and Nature or Physical Geography as modified by Human Action*, New York, 1871. See also David Lowenthal, *George Perkins Marsh: Versatile Vermonter*, New York, 1958.

[4] N. F. Shaler, *Man on the Earth*, New York, 1912.

[5] A. Woeikof, 'De l'influence de l'homme sur la terre', *Annales de Géographie*, 10 (1901), pp. 97–114 and 193–215.

[6] C. J. Glacken, 'Changing Ideas of the Habitable World', in W. L. Thomas (ed.), *Man's Role in Changing the Surface of the Earth*, Chicago, 1956.

for these reasons preferable to the life of towns. In this view of the world, derived from the nineteenth century Romantics, we sense a kindred spirit with Patrick Geddes around 1900, whom Reclus knew in Edinburgh. We suggest also a connection with the later views of Lewis Mumford. Indeed, these three men—Reclus, Geddes and Mumford—are undoubtedly associated with a main trend of this whole period. But let us note in conclusion that these philosophical attitudes, detected in a few pages of Reclus' work, in no way affect the massive labour involved in the mastery of material and clarity of presentation of the nineteen volumes of the *Nouvelle Géographie Universelle*. This great work ranks as a worthy predecessor to the 'universal geography' of Paul Vidal de la Blache.

Élisée Reclus's New Physical Geography was published in 1890 in two volumes, Vol. I: *The Earth* and Vol. II: *The Ocean, Atmosphere and Life*, with the subtitle 'a descriptive history of the phenomena of the life of the globe.'[1] The important point about this work is that it still refers to physical geography as the description of the surface of the earth plus the life of plants, fauna and Man upon it.

> Physical geography, in confining itself to the present epoch, merely describes the earth as it is existing before our eyes. Its aim is not so ambitious as that of geology, which tries to recount the history of our planet during the long succession of ages; but still, it is geography which collects and classes the facts; she it is that discovers the laws both of the formation and the destruction of strata.

Reclus discussed Man's relationship with Nature. This has meant exploration, irrigation, drainage of marshes and lakes, construction of dykes (especially on coasts), and the addition of the varied lines of communication. Reclus pointed out that Man has destroyed natural flora and wild animals and replaced them with his own cultivated crops and domesticated animals. He has changed the balance of Nature, sometimes to his disadvantage, by introducing 'ruptures in the harmony of Nature'. Man has served also as an agent of 'disfigurement and embellishment of the earth'. What is needed is 'a robust education face to face with Nature. . . . This will give us the grandest development of the real love of Nature'.

Reclus' method of regional description may be illustrated from the section on Europe in his 'universal geography' (*The Earth and Its Inhabitants*). It should be compared with the method of Malte-Brun

---

[1] The English translation was done by the anthropologist, A. H. Keane, a fact that indicates the great interest at that time of students of man (notably Keane and A. C. Haddon in Britain) in the progress of geography.

fifty years earlier (p. 20) and Vidal de la Blache fifty years later (p. 218). This deals (as a whole) with boundaries, natural divisions and mountains, maritime regions, climate and inhabitants. France (in particular) is dealt with as follows:[1] position, geology climate and rivers, the prehistoric age, inhabitants (racial characteristics). Then follow general descriptions of the physical features and inhabitants of sections of the country, such as the Pyrenees, *Landes*, and basin of the Garonne. Then follows the topography, in which a description of places, resources, products, and towns of each area are described. Concluding chapters deal with statistics and government. The former covers population (growth and distribution), agriculture (departmental and isopleth maps), mining, manufacturing, commerce, and social data (including a map of the distribution of literacy by departments). The last is a recital of facts of local and central government, education, finance and colonies.

FRANZ SCHRADER

Franz Schrader (1844–1924) had a life-span almost identical with that of Vidal de la Blache, and he was described by Emm. de Margerie as 'le plus original peut-être des géographes français contemporains'. He was not an official, engineer or academic. He early dedicated himself to exploration of the Pyrenees and on the recommendation of Élisée Reclus, and Adolphe Joanne, compiler of famous guide-books, comparable to those of Murray and Baedeker, was employed by the Hachette publishing firm. He was a gifted field surveyor, draughtsman, and artist, and prepared new maps of the Pyrenees, notably of the central section. He became the chief of the *Bureau Cartographique* of the Librairie Hachette, and produced several atlases with the aid of collaborators—*Atlas de Géographie Moderne* (1889), *Atlas de Géographie historique* (1896), *Atlas Universel de Géographie*. The last, prepared in collaboration with Vivien de Saint-Martin and published in parts (1881 to 1911) extending over a period of thirty years is a master piece of cartography. The *Année Cartographique* was also prepared by Schrader and its first fascicule appeared in 1891, the 23rd and last in 1913. It contained a resumé of the annual progress of exploration and as early as 1899 he proposed an international cartographic union.

Schrader wrote many manuals for schools (published by his employers, Hachette) and also gave lectures as a professor at the

[1] *The Earth and Its Inhabitants*, ed. E. G. Ravenstein, Europe, Vol. II, France and Switzerland.

school of anthropology in Paris. One of his last pronouncements was entitled 'The Foundations of Geography in the Twentieth Century'. It was the first memorial lecture in honour of A. J. Herbertson in Britain and was published in 1919. This was the first document on the meaning of geography that I read. Schrader died in 1924 and is buried at the foot of Mont Perdu in the central Pyrenees, in a spot he knew so well.[1]

LUCIEN GALLOIS

Lucien Gallois (1857–1941) was born in Metz and educated in Lyon[2]. He entered the École Normale Supérieure in Paris in 1881 and became one of the first pupils of Vidal de la Blache. At this time the historical tradition was strong in France, through the influence of Himly who preceded Vidal at the Sorbonne. Gallois steadily emancipated himself from this bias in seeking to evaluate the relations between Man and Nature. His doctorate monograph reveals his early historical interest—*Les Géographes Allemands de la Renaissance* (1890), which was an important contribution to the history of cartography. His second thesis (written in Latin according to the regulations of that time) was devoted to the work of a German mathematician-cartographer of the same period. Throughout his life he remained dedicated to cartography and discovery in the sixteenth century and made important contributions.

Gallois, however, early turned to questions of modern geography and became one of its leaders. He spent four years at the University of Lyon (1889–93), where he pursued his first geographical studies in the surrounding areas, which were published in the *Annales de Géographie*. He was called to the Sorbonne in 1893 and soon became associated with the editorship of the *Annales* as assistant to Vidal de la Blache. He wrote many articles for this journal during his editorship, between 1898 and 1919. After a long and active career as researcher, teacher and guide as the director of the Institute for a few years after the death of Vidal, Gallois retired to new tasks in the late twenties. The *Géographie Universelle* was launched by Vidal, but he died before a single volume was published. This remarkable series shares the joint editorial names of Vidal de la Blache and Lucien Gallois and the latter was, in fact, responsible for editing the series

[1] Emm. de Margerie, 'L'oeuvre géographique de Franz Schrader', *Compte Rendu du Congrés Intern. de Géog.*, t. II, (1925), pp. 37–51.
[2] Em. de Martonne on Lucien Gallois, *Annales de Géographie*, 50 (1941), pp. 161–7.

until the time of his death in 1941. Em. de Martonne writes: 'Gallois played an essential part in the development of the French school of geography. He knew how to tackle new lines and to guide the young generation. His name will remain inseparable from that of the master whose work he continued by giving it a precise orientation.'

France remained Gallois' main area of research and his best known work was *Régions naturelles et noms de pays*, published in 1908. De Martonne wrote of this work 'For the precision of its historical and cartographic documentation, and its essentially geographic method and for the decisive clarity of its conclusions, (this work) has never been equalled.' He showed how the idea of the *région naturelle* emerged at the end of the eighteenth century in the preparation of the first geological maps in France. The geologists discovered and used popular names of areas that were indentical with geological units. Gallois carried this enquiry further with special reference to the Paris Basin. As an example, he showed that the pays of Beauce is an area with a platform of horizontal limestone, capped by loess (*limon*), with which there is associated a distinctive landscape of open arable fields and compact villages, the area bearing the name in history and on maps of Beauce. Modern maps reveal villages with the suffix *en Beauce* that were so named in the early nineteenth century to distinguish them from the villages in Beauce with the same name that lay within the newly formed Department of Loir-et Cher. These villages lie on the outer edges of land with the same geological formation as the centre, in a zone of transition to another geological arrangement. Thus Gallois reaches the conclusion that 'It is to the land entities, large or small, all of which are of a physical order, that we should reserve the name of *régions naturelles*. Of the variety and complexity of facts in which the activity of Man intervenes (with Nature), those which bear the influence of the environment form the distinctive field of human geography.' This dictum had a profound influence for the next generation on the purpose and direction of geographic work in France.

EMMANUEL DE MARGERIE

Emmanuel de Margerie (1862–1953) is one of the great figures of modern French scholarship in the natural sciences.[1] He never held a University post nor did he hold a University degree. Yet he soon

[1] Em. de Martonne on Emm. de Margerie *Annales de Géographie*, 62 (1953), p. 389.

emerged as a distinguished geologist and became a life-long advocate of geography. He was restricted by an infirmity from indulging in arduous field work and was above all an insatiable reader, bibliophile, recorder and interpreter. He stands among the pioneers of the study of land-forms, for he served as junior author, with G. de la Noë, in the famous work on *Formes de Terrains*, published in 1888. From 1894 he was associated with Vidal de la Blache in the direction of the *Annales* and submitted to it many distinguished reviews of new works on land-forms at this time including works in German, a language which he commanded with facility. His main labour in this connection was the translation into French, with lengthy annotations, of the classic work of the Viennese geologist Ed. Suess, on *Der Antlitz der Erde* that was originally published between 1883 and 1908. De Margerie's translation included lengthy and invaluable annotations, and was prepared with the assistance of 17 collaborators. It was so highly regarded by Richthofen in Berlin, that he recommended it to his students in preference to the original German edition. De Margerie himself considered his most important work to be his exhaustive geological and morphological study of the Jura that appeared in two volumes, the first in 1922, and the second in 1936. He also produced important geological maps and participated actively with Albrecht Penck in launching the International Map of the World at the International Congress in 1913.

19

The Second Generation:
Pupils of Vidal de la Blache

There occurred a remarkable fluorescence of geography under the direct guidance of Vidal de la Blache for a period of about twenty years, until his death at the beginning of World War I. The tradition he established was vigorously pursued by his immediate successors, notably by Emmanuel de Martonne. No attempt will be made here to list comprehensively the formidable list of his pupils and the works of scholarship they produced, though many of their names appear in these pages. Their weighty contribution to the advancement of knowledge is generally recognized by scholars of all denominations in France. This chapter will be devoted to those followers who became the most distinguished and influential protagonists of *la tradition vidalienne*. They became the leaders of the second generation in the second quarter of this century. These men, all of whom are now dead, are Emmanuel de Martonne, Albert Demangeon, Raoul Blanchard, Maximilien Sorre, and Henri Baulig. André Siegfried must also be considered here. Though he was not strictly speaking a pupil of Vidal de la Blache, his works in political geography reveal the influence of Vidal's thinking.

EMMANUEL DE MARTONNE

Emmanuel de Martonne (1873–1955)[1] was a student of Vidal, and, indeed, married his daughter. He was the recognized leader of French geography from 1918 to 1945.

Born in Brittany in 1873, de Martonne entered the *École Normale Supérieure* in 1892 to study geography under Vidal. He obtained the *agrégation* in history and geography in 1895 (in the same year as

[1] Obituary article by A. Cholley, *Annales de Géographie*, 65 (1956), pp. 1–14. Also note by Jean Gottmann, *Geog. Rev.*, 46 (1956), pp. 277–9.

Demangeon), the doctor of letters in 1902, the doctor of science in 1907. In 1899 he went to the University of Rennes, in 1905 he moved to Lyon, and then four years later he was called to the Sorbonne in Paris to the chair left vacant by the death of his father-in-law, Vidal de la Blache. This post he held until retirement in 1944. He made great achievements for geography in France in furthering the plans of Vidal, including the opening of the large Institute of Geography in 1923. Between the wars it is no exaggeration to say that he was the most influential geographer in Europe. After serving the International Geographical Union as secretary, he became its President. He retained this status during the chaos of the war and emerged thereafter as its first Honorary President. De Martonne, declared his successor A. Cholley, is without question 'une des grandes figures de la géographie moderne'.

De Martonne specialized from the beginning on physical geography and his special area of concentration was central Europe. In contradistinction to many of his contemporaries (and in accordance with Vidal's teaching) he had a firm grounding in geology, geophysics and biology. His first published work dealt with the coastal morphology of Brittany and he soon turned to studies of physical geography in the Carpathians. His regional monograph was on Wallachia and appeared in 1902. He maintained a life-long interest in glacial erosion in the Alps, beginning with a paper in 1911, although his interpretations were always contended by Raoul Blanchard (see below). Among the major works of de Martonne, special mention should be made of his *Traité de Géographie Physique*. The first edition appeared in 1909 as one volume of 190 pages, though it was extended to three volumes in 1925–7. Many studies, by himself and his students, culminated in *La France Physique* in the *Géographie Universelle* in 1942. While primarily a physical geographer, de Martonne, like Henri Baulig (below) sought to combine the physical and human aspects in one framework. This is evident in the two volumes on Central Europe in the *Géographie Universelle* (1930 and 1931).

De Martonne first visited the United States in 1904 for the eighth International Geographical Congress and was at that time much impressed by W. M. Davis' work on geomorphology. He returned for the transcontinental excursion that was organized by Davis in 1912, and received the American Geographical Society's Cullum medal in 1939. He last visited New York in 1949 as head of the French delegation to the UNESCO conference on Conservation and Utilization of Resources. He was a firm friend and colleague of several American geographers and worked closely with Bowman during the Paris Peace Conference in 1919. Gottman informs us that

The Second Generation: Pupils of Vidal de la Blache

de Martonne was 'an enthusiastic follower of Davis and a great admirer of Johnson and Atwood; these names were classics to de Martonne's students'. He edited the *Annales de Géographie* and the *Atlas de France*, managed the International Congress in Paris in 1931, was its secretary from 1931 to 1938, and president from 1938 to 1949. Gottman concludes: 'He lived for his work and inspired many others to follow in the same path.' Such are the true attributes of a leader and maker of his discipline.

ALBERT DEMANGEON

Albert Demangeon (1872–1940),[1] whose parents came from the Vosges, studied under Vidal de la Blache at the *École Normale Supérieure* in Paris where he achieved the *agrégation* in history and geography in 1895. He was a contemporary student, friend and colleague of de Martonne. At first he taught in several secondary schools in Picardy. It was during this period that he undertook studies that culminated in a monograph, published in 1905, with the title of *La Picardie et les régions voisines*. As a young student of Vidal de la Blache, he achieved in this book what de Martonne describes as 'reasoned description, complete and vivid, of a region of France', and this study was the first of its kind. On the strength of it he was called to the University of Lille where he remained until 1911 when he was called to the chair of geography, alongside de Martonne, at the Sorbonne in Paris. Here he remained until his death in 1940.

While at Lille, Demangeon commenced studies in the Limousin, which, though not completed, were published as two important articles in the *Annales*. One of these articles was on the human use of the *montagne*, the other was on the relief and its development through a chequered series of cycles of erosion, rather than one uninterrupted process as postulated by W. M. Davis. In fact, Davis was so interested in this work that he visited the area together with Demangeon—still another indication of the impact of Davis' ideas on European workers at this time.

Demangeon, however, concentrated on problems of human geography. His work on Picardy demanded familiarity with the archives in the national library in Paris. This topic was the subject of his complementary thesis and was also published in 1905, and the work has received general recognition from historians and others con-

[1] Em. de Martonne, 'Albert Demangeon', *Annales de Géographie*, 49 (1940), pp. 161–9.

cerned with such documentary work. He also developed the technique of the questionaire that he pursued through his professional life and many others have followed.

When Demangeon moved to the Sorbonne, at the age of 39, he continued to specialize on human geography. He fixed his sights on writing a manual of human geography and an economic geography of France. In thirty years he completed neither under these specific headings, but he achieved much else. The *Géographie Universelle* had among its first contributions Demangeon's classic work on the British Isles (translated in English) and also on the Low Countries, (Netherlands and Belgium). These works appeared in the mid-twenties. His interest in Britain was revealed in a work on its 'colonial geography', published in 1923 (translated and published in England in 1925, in Germany in 1926). His *Declin de l'Europe*, published in 1920, was the assessment of the losses of World War I and the potentialities of recuperation thereafter.

He devoted much of his time to the editorship of the *Annales*, and contributed to this journal 31 articles and 89 notes. These included *Habitation rurale en France* (1920), a study of the regional variations of the farmstead, which he pursued throughout his life; and *L'Habitat rurale* (1927), a pioneer study on the distribution of the varied forms of nucleated settlement in rural areas over the earth. He was the prime mover in establishing a research commission of the International Geographic Union on this theme, and thereby in stimulating such research in many lands.

Demangeon also turned to the geography of transport, of overpopulation and of international economics. In the thirties he became further involved in the preparation of a "questionnaire" on rural farmsteads, agrarian structures, and the distribution of foreigners in agriculture in France. This work received the support of the Rockefeller Foundation, and was based on an elaborate questionnaire that was distributed to knowledgeable persons in numerous communes throughout France. He revised his classification of farmsteads and produced remarkable maps of their distribution that are reproduced in the work on France in the *Géographie Universelle*. This field of enquiry has always been regarded in Britain and the United States as *avant garde*, or of questionable concern to geographers, but it has been recognized for over fifty years in both France and Germany. Demangeon's ambitions, expressed on arriving in Paris in 1911, were not realized, but the materials and ideas are incorporated in his magnificent two volumes on France published in the *Géographie Universelle*. He had just checked the proofs of these works when he died suddenly in 1940.

The Second Generation: Pupils of Vidal de la Blache

Demangeon never wrote the definite treatise on human geography that he envisaged throughout his life. But the memorial volume on 'problems of human geography', being a posthumous collection of his writings from 1902 to 1941, reveals clearly the nature of his conceptual framework. He envisaged human geography as the study of human groups in their relations to the geographical milieu— not an abstracted 'physical environment', as was usually assumed in the early years of this century, but the transformed milieu of man's creation in which he works and has his being. Thus, it has four fields of enquiry—the influence of the natural milieu on modes of life (*genres de vie*)—one of the primary ideas of Vidal de la Blache a generation earlier; the changes in these modes of life under the impact of the progress of human societies, or, in other words, the influence of the human milieu; the distribution of human groups as the result of the natural milieu and the 'degree of civilization', or, as James would express it, the attitudes, objectives and technical abilities of the human groups; and the establishments (works) in the landscape, due to the impact of the human group on the land, or more precisely sites, such as are occupied by the house, the village, the field boundary, the track, the village, and the most complex groups, the city and state. In these fields of enquiry Demangeon invokes three guiding principles. The first is the Vidalian idea of 'possibilities' of human occupance of land rather than determinants of human occupance by physical site conditions; the second, never to lose sight of the territorial substratum (land), and build up always inductively from a 'regional base'; and the third, to recognize always the fundamental importance of historical perspective so as to explain the present modes of human occupance of the earth.

Jean Gottmann, who was Demangeon's last personal assistant (from 1936–40) writes as follows in a letter to me.

One day during the winter of 1932–33, the first year I was a direct student of Demangeon and attended his seminar at the École Normale Supérieure (among the Normaliens in the seminar that year was Pompidou, the present French prime minister), I had a long talk, eye to eye, with Demangeon about the future orientation of my work. The old man told me that he felt the younger generation of geographers should go into the study of one of four major problems which were going to be especially important in the future, in his opinion. His own generation had not given much time to study these problems.

1. the crowded masses of the Far East. He had put one of his better students, Etienne Dennery, on this trail, but he was

shifting towards diplomacy (Dennery retired last year as ambassador to Japan and is now head of the Bibliothèque Nationale). But now Gourou and Robequain were working in this area.

2. the relations of the whites with the Negroes. He had put Jacques Weulersse in this field, but J. W. had shifted to the Middle East and was working in Damascus.

3. Irrigation in the arid countries. There were a whole string of students engaged in that direction.

4. the growth of great cities.

Demangeon wanted me to specialize on the supply and markets of Paris (beginning with dairy products). I was reticent and fought for something more exotic to win an assignment and a trip to study irrigation in Palestine. In 1963 I remembered the old man's forecast when I was rather proud to be called upon by the metropolitan authorities to chair a debate in Paris on the future of 'Les Halles'.

Not many in our profession have had the insight and the foresight of Demangeon. He was a great teacher though a modest man, and it is difficult to assess the evolution of the French school in the tradition of Vidal without accounting for his part.

RAOUL BLANCHARD

Raoul Blanchard (1877–1965)[1] was in the fullest sense a maker of modern geography. 'Some geographers spend their lives trying to define geography. As for myself, I make it'. These are Blanchard's own words reported by Raymond E. Crist. He was born at Orleans in 1877. He passed the formidable *agrégation* at the age of 23 (1900) in Paris, where he studied under Vidal de la Blache.

He first spent several years at Douai and Lille and completed in 1906 his doctorate monograph on Flanders. On the strength of this he was called to, and accepted, the chair of geography at Grenoble and there he spent almost fifty years. Here he established a school of geography in the *Institut de Géographie Alpine* in 1908 and a journal entitled *Revue de Géographie Alpine*. He commenced his own great work on the French Alps in 1937 at the age of sixty and continued work on it until the publication of the final and twelfth volume in 1956. In 1917 he first visited Harvard as an exchange professor and thereafter produced much work on that continent, more especi-

[1] Jean Dresch and Pierre George, 'Raoul Blanchard', *Annales de Géographie*, 75 (1966), pp. 1–5. Also a note by R. E. Crist in *Geog. Rev.*, 55 (1965), pp. 602–3.

ally on eastern Canada. He subsequently visited Columbia (1922), Chicago (1927), Harvard (1928–36), Berkeley (1932), Montreal (1947), and Laval (1952), and in 1956 he was awarded the Charles P. Daly medal of the American Geographical Society and in 1960 the gold medal of the *Centre National de Recherche Scientifique*, the French equivalent, as Crist reminds us, of the Nobel prize. Blanchard inspired and communed with numerous students at his institute, and some dozen institutes in other Universities are headed by his pupils. His institute in Grenoble was in more senses than one a pole of research and geographical scholarship independent of Paris. His views and researches kept pace with the times, but his basic terms of reference as a geographer remained constant through his lifetime and received universal recognition. 'The work of Professor Raoul Blanchard, Member of the Institute, belongs to the ages' (Crist).

Blanchard was one of the pupils and followers of Vidal. He produced a regional monograph in 1906 on Flanders (*La Flandre: Etude Géographique de la Plaine Flamande en France, Belgique et Hollande*). His first paper on rainfall in northern France was published in the *Annales de Géographie* in 1902. In his fourth year at Grenoble he produced a book on the geography of Grenoble (1910) that stands as a classic, and has been through a number of revisions; and it is incorporated, with new themes, in his great work on the French Alps. He also produced two substantial volumes on the geography of eastern Canada containing long geographies of the cities of Quebec and Montreal.

The general purpose of Blanchard's work as a geographer was 'the description of the environment of man and his activities'. His method was based on personal observations and recordings, interviews and enquiries. To all this I can testify from two weeks alone in the Bauges with *le géographe alpin* when he was preparing that section of his first volume on the Alps in 1937. But while Blanchard's work was focussed on the field, he, like Demangeon, also used the materials of the historian and archivist. He also found time to write general text books on North America and Europe, and was part author of the volume on western Asia in the *Géographie Universelle*, published in 1927. Particularly important, however, in the aggregate of his life's work were his studies of the French Alps and eastern Canada.

Among his personal researches, outstanding merit attaches to his work on the morphological problems of the Alps and on urban geography. He opposed the morphological cycle of de Martonne in its relevance to the Alps (which had close affinities with the views of Davis) by pointing out the role of structural and lithological conditions and glacial action. This always remained a bone of contention between de Martonne and Blanchard. To the study of urban geography he made his classic contribution on Grenoble in 1910—an exemplary study among a number of others that appeared from the Vidal school in the 1900's. Blanchard later continued this work in studies of Annecy, Quebec and Montreal. After the Second World War, he took advantage of his own and other researches, and tackled new urban problems, which were implicit but unanswerable in his original study of Grenoble e.g. the role of the regional capital, and migration of population to the city. This is also evident in his section on Grenoble in the volume on the French Alps, and his latest studies on Annecy in 1957 and Nice in 1960. In the fullest sense one can speak of Blanchard's *École de Grenoble*, and this he always envisaged as a pole to the other major focus in Paris.

Blanchard established at Grenoble a school of geography independent of Paris and his pupils between the wars found places, and most of them are still active, at various institutions throughout France. Among his outstanding doctoral students (out of a total of 15) are the following with the titles of their theses—Bénévent, *Le Climat des Alpes Françaises*, 1926; Jules Blache, *Le Massif de la Grande Chartreuse et du Vercors*, 1931; A. Allix, *L'Oisans*, 1929; M. Pardé, *Le Régime du Rhone*, 1924; Ph. Arbos, *La Vie Pastorale dans les Alpes Françaises*, 1922; H. Onde, *La Maurienne et la Tarentaise*, 1931; R. Robequain, *Le Thanh Hoa: Étude géographique d'une province annamite*, 1929; C. P. Peguy, *Haute Durance et Ubaye* 1947; and G. Veyret-Vernier, *L'Industrie des Alpes Françaises*, 1948.

MAXIMILIEN SORRE

Maximilien Sorre (1880–1962)[1] began his career as a school teacher and, in this capacity, completed his thesis in Montpellier in 1913. This was prepared under the auspices of the plant geographer, Charles Flahault, although his thesis on the Mediterranean Pyrenees was essentially based on the concept of the *genre de vie* as put forward by Vidal de la Blache. The slant of this work towards a biological (ecological) approach gives at once the key note to Sorre's life work. He entered military service in 1914, was seriously wounded and demobilized in 1915. He resumed 'enseignement supérieur' at Bordeaux and then at Strasbourg, where he co-operated in educa-

[1] P. George, 'La Vie et L'oeuvre de Max Sorre', *Annales de Géographie*, 71 (1962), pp. 449–59 with portrait and bibliography.

tional work with Henri Baulig, Lucien Febvre (the historian and author of the *Geographical Introduction to History*, 1924), and Marc Bloch, the distinguished medievalist. He was later located for nine years in the Faculty of Letters in the University of Lille. He became increasingly involved in administrative work for which he had a rare talent and was appointed a Rector of the University at Clermont Ferrand and then at Aix-Marseille in 1934. In 1940 he succeeded Albert Demangeon to the chair of geography in Paris and there he remained until his retirement.

Sorre remained dedicated throughout his life to the geography of man. His first small publication dates from 1904. The culmination of his work is the four volume opus on *Les Fondements de la Géographie Humaine* (1947–52), a work which finally does for human geography what de Martonne's work did for physical geography fifty years before. It will long remain a classic and deserves a full translation into English. His last single work on *l'Homme sur la Terre: Traité de Géographie humaine*, published in 1961, contains new ideas, but it is essentially a resumé of the larger work. The approach is throughout ecological and provides a basic conceptual framework for the geographer and innumerable stimuli for future research.

Special attention should be given to the fundamental work of M. Sorre on the foundations of human geography. The first volume deals with the biological foundations of Man's use of, and life on, the earth. This embraces climatic adaptations; the distribution and spread of domesticated plants and animals; human nutritional needs and the distribution of human diets; and pathogenic complexes, that is, associations of physical conditions by which the distribution of human diseases may be located and explained. The second volume deals with the technical foundations of human occupance of the earth and the development of communication and embraces the technics of social life, i.e. organizations beyond the family. Social technics embrace language, clans and tribes, local sedentary groups, nations, states and empires. The economic technics include energy and production and their localization, in terms of animate and inanimate sources of power (including human muscle) and natural resources. The development of communications deals with traffic by land, water and air and the geographic impact of circulation, that is, the patterns and causes of human movements and exchanges. The third volume deals with the technical foundations of production and the transformation of raw materials—exploitation of the animal realm, forestry, the use of water, soil fertilization, systems of cultivation and stocktaking, minerals, and raw materials. The fourth volume embraces 'the human habitat'. Sorre takes off from Vidal's concept of the *genre de vie* that embraces the totality of activities of the human group in relation to environment, and owes its character to particular combinations of technics. The rural habitat deals with the mode of grouping and distribution of agricultural population and those who service them, and is distinguished from the rural habitation, or, as it is called in our land, the farmstead and the working quarters. The urban habitat is examined in terms of the urban functions, morphology, metropolitan status, physical structure, population, and the impact of the urban centre on its surrounding countryside. This is a mere skeleton outline of the arrangement of this voluminous work. Every student of geography needs to digest the relevant sections as a springboard for his own work.

These works reflect Sorre's life-long preoccupation with the fields of medicine and sociology. He was a voracious reader and was fully conversant with the works of past and current geographic scholarship in Germany. For Sorre physical geography was a framework for the study of human groups and he was concerned with the physical environment only in its relevance to human occupance.

HENRI BAULIG

Henri Baulig (1877–1962) began as a student of Vidal de la Blache,[1] and in fact, he was for some time Vidal's assistant. At the suggestion of Vidal he gave up this post and the intention of preparing for the *agrégation*, and left for the United States in 1905. He there came under the strong influence of Davis at Harvard and from then on maintained a strong professional interest in the field of geomorphology and in the geography of the United States. Back in France he first returned to the Sorbonne and in 1913 moved to the chair at Rennes. He served in the First War, then moved to Strasbourg where he remained until his retirement in 1947. The Second World War interrupted his work, in a sense, for his University was transferred to Clermont Ferrand, and through his active interest in subversive activities against the Vichy regime he was arrested in 1944. After the war, he returned to Strasbourg but retired in 1947. He continued his active professional work for the next ten years and died in 1962 at the age of eighty-five.

Henri Baulig shared with de Martonne the distinction of being a leader of geomorphological research. His first work was on the morphology of the Central Massif. Taking the Davisian framework

[1] An obituary article on Henri Baulig, *Annales de Géographie*, 71, (1962), pp. 561–6.

as a springboard (and he was a close friend of both Davis and Johnson), he looked at the solid surface of the earth—in other words, he sought to determine the varied processes involved in the emergence of landscape not merely those processes which Davis theoretically isolated. Baulig sought the *explication d'ensemble de l'organization du relief terrestre.*

In his 72nd year his collected works were assembled in one volume as *Essais de Morphologie* (1950) and several years later he wrote two long articles in the *Annales de Géographie* that were a critique of the specialized methods and claims and trends of the new geomorphology in France as expounded in the major work of P. Birot.[1] Baulig had a remarkable gift for observing and interpreting physical land-forms and his work on the role of eustatic equilibrium in the formation of the Central Massif is one of rare distinction. Yet he always retained the geographic outlook according to the tradition of Vidal de la Blache, as did his colleague de Martonne. His work in the *Géographie Universelle* on North America is probably the finest geographical work in print of that continent and he received the Daly award of the American Geographical Society specifically in recognition of this signal contribution. He continued to give attention to the field of geography in general (as well as to questions of geomorphology) after retirement and he welcomed and launched an atlas of eastern France in 1960.

ANDRÉ SIEGFRIED

André Siegfried (1875–1959)[2] was not a direct student of Vidal de la Blache in the sense of taking courses with him, but he saw him often and Vidal was one of the inspirers of his early work. However, to Vidal's human geography Siegfried, from his beginning, wanted to add political geography, and this was largely inspired by the tradition of Tocqueville, as witnessed by the title of the first of Siegfried's books and also his doctoral dissertation, *La Démocratie en Nouvelle Zélande* (1904), which was obviously a renaissance of the tradition of *La Démocratie en Amerique*, a book which had been neglected in France in the nineteenth century although highly influential in America. An important work in political geography was his study of political attitudes in western France (1913). After Vidal's death Siegfried kept in close touch with Demangeon, and they saw each

---

[1] P. Birot, *Les méthodes de la morphologie*, Paris, 1955, Reviewed by Henri Baulig, *Annales de Géographie*, 66 (1957), pp. 97–124 and 221–35.

[2] These comments on Siegfried are based on a letter from Jean Gottmann.

other often though Siegfried was of a different level of French life and society from the regular geographer. He belonged to one of the leading families of French protestantism, which had many family and business ties with the high society of North America, Britain, Canada etc. To understand his training, you must read the little book he wrote in the 1950's about his father Jules Siegfried, who was the economic and political boss of Le Havre, a personal friend of Abraham Lincoln and other American presidents, several times a cabinet member in France and one of the leaders of the cotton textile industry. André Siegfried was a cousin of so many big bankers, politicians and statesmen, that his means of information and his audience were of another quality than that of other geographers, who obviously resented him. They said he was not a geographer but a mixture of economist and political scientist. But Siegfried always considered himself a geographer. In his youth he tried to enter a political career but twice failed to be elected to Parliament; then, after 1910, he chose to specialize in analysis of the political process through the geographical method. At the school of political science in Paris (of which his father had been one of the founders), he called his courses Geography and he helped younger geographers to gain recognition on the faculty (including myself). When he was elected to the Collège de France, he chose to call his chair 'Géographie économique et politique', which had never existed before. On his retirement, he saw to it that a geographer succeeded him (Pierre Gourou in 1947). However, Siegfried remained a wonderful analyst and student of special cases and never produced a general theoretical system, either in politics or in human geography. He inspired many younger people, either among geographers or among political scientists, and in many ways he created a school of thought, though not formally so.

20

The Third Generation

Among the many senior geographers of the third generation five represent distinct schools of thought, and have emerged directly from the foundations laid by Vidal de la Blache. These are Pierre Deffontaines, Roger Dion, Pierre George, Jean Gottman and André Cholley.

PIERRE DEFFONTAINES

Pierre Deffontaines (1894–), born at Limoges, took his first doctorate in law at the University of Poitiers in 1918. He then proceeded to Paris to prepare for the *licence* in geography and history, for which he qualified in 1919. His first teachers were de Martonne, Demangeon and Gallois. A further diploma in *études supérieures* was awarded in 1921, when Demangeon was chairman of the examining committee. The subject of the dissertation for the diploma was the 'prehistoric geography of the Limousin', which, as Deffontaines claims in a personal letter, was the first study of its kind in France.

His contact with Jean Brunhes began in 1918, following an enthusiastic reading of the master's work on human geography. Deffontaines writes that he submitted to Brunhes an essay on types of habitation in the Basque country. Brunhes was greatly interested in this and asked Deffontaines to collaborate in writing a work on the human geography of France. They prepared together a map of roof types in France which is now well known. Deffontaines writes that 'Brunhes was an astonishing master, continually reaching out to new ideas and constantly discovering new horizons of work'.

Jean Brunhes was influential in getting for Deffontaines an award from the *Foundation Thiers* in 1922. This he held for three years, at the same time as another outstanding geographer, Roger Dion. Deffontaines took the *agrégation* in history and geography in 1922 and his doctorate thesis was completed and submitted at the Sorbonne in 1932, with Demangeon as chairman, and de Martonne and Gallois as co-examiners. The subject of the thesis, published in 1932 and discussed in Chapter 17, bore the title *Les Hommes et leurs Travaux dans les pays de la Moyenne Garonne*. A smaller accompanying work, entitled *La Vie forestière en Slovaquie*, was published by the *Institut d'études slaves de Paris* in the same year. The former work is remarkable for its originality of conception and arrangement and was a radical departure from the approach of other monographs. The distinctive character of this work is described as follows in Deffontaines' own words in a personal letter, dated December 9th, 1965.

The originality of my thesis on the middle Garonne is, in the first place, that it was the first thesis on an area south of the Loire. It was above all oriented in order to show, after the manner of Brunhes, that the density of population is dominated by the density of the horizons of work (*horizons de travail*). Reduction of population comes mainly from a reduction of the opportunities for work. I looked back in history by parting from the present, for I believe that the geographer is concerned with the past only in so far as it explains the present. The region (*sic*) I studied was not a natural unit, but, on the contrary lies at the point of meeting of all sorts of influences from surrounding areas. The thesis begins with an album of photographs with a long commentary on each of them. This serves as an introductory presentation of the area before proceeding to the detailed study.

Right through the twenties Deffontaines' career reveals a keen interest in prehistory and ethnology. In 1926 he was the secretary of the International Congress in Anthropology at Prague. In 1934 he was invited to the University of Sao Paulo in Brazil to establish a new chair of geography. Shortly afterwards he was invited to the Federal University of Rio de Janeiro, where he taught from 1936 to 1938, founding during this short period the National Committee on the Geography of Brazil and a new geographical periodical. In 1939 he left Brazil to become director of the *Institut Français* at Barcelona, Spain. He has travelled widely since the war in South America and received an honorary doctorate degree at the Federal University of Rio de Janeiro in 1946. In 1952 he was invited to Laval University at Quebec, Canada, to launch a new department of geography, here again he received an honorary doctorate and was invited to return there in the summer of 1967. In 1965 (at the age of 71) he was invited to become Professor of Geography at the University of Montpellier, which he accepted after 25 years at Barcelona.

The work of Pierre Deffontaines reveals the continuation of the interests and approach of Jean Brunhes. While he was working on his doctorate work in the late twenties, he found time to collaborate with Brunhes in writing the second volume on the human geography of France (600 pages), published in 1928, in the series of volumes on *L'Histoire de La Nation Française*, edited by Gabriel Hanotaux. (The first volume was written by Brunhes.) He has edited a series of books on *Géographie Humaine* of which he has written volumes on colonization, man and the forest, the geography of religions, and man and the winter (in Canada). He has also directed a *Géographie Universelle Larousse* in three large volumes (1959–60) and a *Géographie Générale* in the *Encyclopadie de la Pléiade*, in collaboration with André Journaux and Mme. Jean Brunhes Delamarre (1966). The last named lady is the daughter of Jean Brunhes, and she and Deffontaines have revised jointly Jean Brunhes' work on *La Géographie Humaine*. This appeared in a new edition in 1934, and in an abridged form in 1942, of which a second edition appeared in 1947. The keen interest in human geography and ethnology of Deffontaines runs through all his work, and is evidenced by the foundation of a new periodical *Géographie Humaine et Ethnologie* in 1948, which unfortunately ceased publication after the appearance of four numbers. Deffontaines is the most able and prolific disciple of Jean Brunhes.

ROGER DION

Roger Dion (1896–   ) went to school at Blois and served in the army throughout the First World War. Thereafter he studied at the *École Normale Supérieure* from 1919 to 1922 and took the *agrégation* in history and geography in 1921 at the age of 25. Like his contemporaries, his studies were directed, and his examinations supervised, by the immediate successors of Vidal de la Blache. From 1922 to 1924 he engaged in full-time research, supported like his fellow-student, Pierre Deffontaines, by the *Foundation Thiers*. Between 1924 and 1934 he worked on his thesis, *Le Val de Loire*, for his doctorate in letters, a work that is discussed on p. 218. From 1934 to 1945 he was at the University of Lille, first as *maître de conférences*, and then as professor, and from 1945 to 1948 he was Professor of Political and Economic geography at the Sorbonne, and, over the same period, was in charge of an introductory course to the human geography of France at the *Ecole Normale Supérieure*. In 1947 he served one session as Professor at the University of Sao Paulo in Brazil, where there has been a long and close connection with the geographers of Paris. Since 1948, that is for the past twenty years, he has been a research professor in the historical geography of France at the *Collège de France*.

The appointment he has now held for so long is a recognition of the field of scholarship which Dion has made so distinctively his own, though it is one of concern to many French scholars in various fields. This specialization and its standard of excellence is revealed in his extended monograph on *Le Val de Loire*. His remarkable command of skills in the interpretation of documentary records, as was so rigorously counselled by his predecessor, Albert Demangeon, is evident in his works, in which such records are interpreted in the work of Man on the land in creating his habitat. The next work, after the thesis, was a remarkable essay, generalized from case studies of documentary records in different places throughout the land, on the formation of the rural landscapes of France (1934). He traces the differential development of agrarian systems in northern and southern France, and their role in accounting for difference in the emergence and present forms of field patterns, road nets, and the grouping of farmsteads and villages. He has since worked for many years on the geography of viticulture in France and a major work, published in 1959 and covering nearly 800 pages, deals with this theme from the earliest times to the nineteenth century. In 1961 a work was published on the history of *levées* on the Loire (312 pages), an obvious outgrowth from the subject of his earlier doctoral thesis. He has also more recently undertaken research on ancient roads and town sites. Latterly (in the sixties) he has turned to topics of the classical and Roman periods, notably to the voyages of Pytheas, and to certain of the Roman roads. As a geographer, Dion exemplifies the competence in historical skills, that is essential to understand the origins of both countryside and town. The depth and patience of his scholarship are reflected in his dedication through his lifetime to enquiry in his native land. He reflects the continuity of the historical tradition in geographic scholarship in France.

PIERRE GEORGE

Pierre George (1909–   ) has been Professor of Geography at the Sorbonne since 1948, and is one of the most prolific writers and 'popularizers' among geographers. He has written or edited many volumes in the *Que sais-je?* and *Magellan* series for the general public, various works as texts for the student, and several other works that are directed to the small specialist market. These professional works include books on the geography of cities, the countryside, economic geography, urban geography, and rural geography, all published within the last ten years. George has also produced substantial researches, which will be noted shortly.

There is a remarkable contrast between the works of Dion and George. Dion is concerned with the past and remains essentially esoteric in interest and his works reach only a small circle of kindred specialists. He has not the least interest, to judge from his publications, in the general audience. George, on the other hand, is very much concerned, beyond his particular researches, not only with advancing the status of geography among scholars, but also in disseminating its approach and contributions among the largest possible public.

It is highly significant to note that George in his training, and by printed acknowledgement, builds upon the conceptual framework of his two main teachers, Albert Demangeon and Maximilien Sorre. In other words, his approach is ecological and he gives strong emphasis to the *motifs* of characterization, action and organization of communities as spatial complexes. I suspect that, like Sorre and Le Lannou, he would delete the adjective 'human' from geography, and would thereby seek to eliminate the increasing gulf that separates the 'physical' from the 'human' aspects of geography.

Pierre George studied at the Sorbonne from 1926 to 1930, taking his *licence* in history and geography on the arts side, and geology, mineralogy and botany on the science side. He took the *agrégation* in history and geography in 1930 and his doctorate was taken in 1935 with the main thesis published as *La région du Bas-Rhône: Étude de géographie régionale* and the subordinate thesis, published as *La fôret de Bercée* (1936). He continued with a work on Bas-Languedoc, published in 1938. (*Études géographiques sur le Bas-Languedoc*). He began his career as a teacher, but became professor at the University of Lille from 1946 to 1948 and in the latter year he was appointed professor at the Sorbonne and has been there ever since.

In a new little book on *Sociologie et Géographie* (1966) George writes that geographers in France habitually claim the unity of geography 'to appease their consciences', but, he asks, does that unity really exist as geography has been practised by the descendents of Vidal de la Blache? His answer to this question is no. What relation, he asks, does the skilled analysis of natural processes, pursued in the laboratory or the field, have to the understanding of geography as a *science humaine*, which, he states categorically, is the traditional French attitude. The explanation of physical phenomena (continues George) is the domain of the natural sciences, a fact that is recognized, for example, in the Faculty (much bigger than a 'department' or a school, be it noted) of Geography in Soviet Universities. On the other hand, in the same country, the economic sciences are dedicated to the adaptation of human action to the conditions offered by the land (*paysage*) which is the synthesis of the studies of various natural sciences. This, incidentally, is the approach (as stated by Journaux and quoted on p. 221) throughout the arrangement of the recent volume on *Géographie Générale*.

The central concern of geography, still according to George, is with the study of human societies and this, he asserts, is the essence of the French approach to geography as formulated by Vidal de la Blache. As a 'human science' its object is 'the study of the globe as a whole and of its segments, with respect to all that conditions and all that is relevant to the diverse human collectivities that make up the population of the globe'. Le Lannou in a work published in 1949 (*La Géographie Humaine*) wrote similarly of *homme-habitant*. He uses this term in order to make clear that man's environment in space involves two aspects, the abstracted natural environment, and the inherited cultural environment, both of which vary spatially and which it is the geographer's special task to characterize and evaluate. (See Bobek's similar emphasis of the cultural environment and his map of types of such environments, reproduced in Marvin and Mikesell, *Readings in Cultural Geography*, 1962.) Commenting on Le Lannou, George prefers to substitute the term *homme-producteur*, or, even better, he adds, *homme-consumateur*. He emphasises the economic base and organization of human activity of the earth on what is relevant to Man. This, of course, depends essentially on the policy and philosophy of the State and is fundamentally involved with national policy. The task of geography, as a scientific study, asserts George, involves what he calls *horizontal* synthesis, rather than *vertical* analysis. Its basic concern, therefore, is with the location and measurement of types of correlation of spatially arranged phenomena over the face of the earth.

JEAN GOTTMANN

Jean Gottmann (1915–), though still relatively young, has achieved recognition as a leading scholar in France and is also widely known in the States for his publications in English. He is professor at the *École des Hautes Études* at the Sorbonne and at the *Institut d'Études Politiques de l'Université de Paris*. He was appointed to the chair of geography in the University of Oxford, England, in

1967. Born in Russia, he went to school in Paris and attended the Sorbonne. From 1936 to 1940 he was an assistant to Demangeon, and since 1941 has divided his time and researches between France and the United States. He states that his 'teachers in the broad sense' were Demangeon, de Martonne, E. Gautier, Jules Sion and André Siegfried, and scholars at the School of Advanced Studies in the States, notably Robert Oppenheimer. Gottmann has been invited many times to the United States (the last in 1965) as a member of the Institute of Advanced Studies at Princeton, N.J. He has also lectured as a visiting professor at various Universities, particularly Princeton, Johns Hopkins, Columbia, Pittsburgh, Southern Illinois, and California (Berkeley). He has also given lectures in Universities in Britain and Canada.

After initially specializing on the human geography of France and the Mediterranean countries, he turned after the Second World War to the general theory of economic and political geography. These studies had particular reference to the countries of 'advanced civilization', or, as he writes, 'over-developed countries' in North America and Western Europe. Since the rapid urbanization of our epoch has made *le fait urbain* particularly crucial in these countries, Gottmann has devoted most of his work since 1950 to questions of urban growth and regional management.

Since 1940 Gottmann has been a consultant on various occasions to economic and planning services in France and America. In 1946–7 he directed research at the Secretariat of the United Nations at New York. From 1949 to 1952 he was chairman of the commission on regional planning of the International Union, and in 1965 he became a member of the Committee on Resources and Man of the National Academy of Sciences. He is a member of many learned societies and a recipient of high merit awards from the *Societé de Géographie de Paris* and from the American Geographical Society.

Publications embrace about a dozen large volumes and over a hundred research articles. Reference is made here only to the books. The first was on the commercial relations of France (1942). A text on America in French appeared in 1949 and another on the geography of Europe in English, was published in New York in 1950 (2nd ed. 1962). A book on the political geography of States (*La Politique des États et leur Géographie*) appeared in 1952. He edited a small collaborative work on regional planning for the International Geographical Union in 1952. The two major works of Gottmann, both of which were commissioned and published in the States, are *Virginia at mid-century* (1955) and *Megalopolis* (1961). He has recently had published a series of essays on the 'planning of occupied land' (*Essais sur l'amènagement de l'espace habité*) published in 1966. A paperback was published in 1966 (*Metropolis on the Move*), containing essays on the urban explosion that were delivered at a colloquium in honour of Gottmann at the University of Southern Illinois in 1963.

The name of Jean Gottmann is well known in America, where he is reckoned undoubtedly as one of the world's leading contemporary geographers. It is a tribute to him and to the geographic tradition of France, that he should be invited in America to undertake important tasks of compilation and presentation rather than calling on the services of an American scholar. This is not the place to examine his works, but the ideas and his approach are very germane to the theme of this book, for they have had widespread repercussions, and they reveal to the English speaking reader the mode of approach of the continental European geographer. I refer particularly to the three major works published in the States. The *Geography of Europe* is an eminently readable work, which I have used as a text on many occasions, since it reveals the French approach to geography to the American student and scholar, so many of whom are currently allergic to the so-called 'regional approach'. Gottmann was invited to prepare a volume on the State of Virginia with the same approach as the text. The result is a work that is, in effect, close to the treatment of the 'regional monograph' in the French tradition. The third major work on the urbanized area of the Atlantic seaboard, that Gottmann called Megalopolis, was undertaken at the request, and with the support, of the Twentieth Century Fund in New York, whose policy it is to publish studies on national and international issues that shall be both authoritative and readable. That a Frenchman should have acquired such remarkable command of the English language may well be overlooked. The book on Megalopolis reveals a viewpoint and procedure that is distinctly geographical in its approach and expertise in its examination of the spatial impact of urbanization in the northeastern seaboard and its impacts on human living conditions and traffic problems. The fact that the whole of this volume has been printed (1964) as a paperback by the press of the Massachusetts Institute of Technology is proof of the demand and prestige of this work.

The philosophy and procedure of Gottmann are revealed in a short article on political geography in the book on *Géographie Générale*, published in 1966, edited by Deffontaines and Journaux. The fundamental aspects of political geography are succinctly described as *cohabitation* (the clustering of interdependent units at particular points of settlement); *répartition* (the modes and methods of distribution necessary to the clustering); *réseau des accès* (the transport network, that results from the modes of clustering and distribution, both past and present). This summarizes the essentials that Gottmann has sought to apply in his substantive published works on man's occupance and organization of space. He writes that 'the circulation of people, of goods and ideas' modifies and animates human settlement and the value of a place depends on its access to the spatial system in which it is located. The control of circulation has always been, he says, a primary aim of political authority and the mode of operation of this authority is described as the 'iconography of the group', or, in other words, the social and political philosophy and action of the State. 'Circulation and iconography have always been associated in the formation of public authority and in the political life of peoples.' Thus, the main factors to be considered in political geography, when regarded as the impact of political action on the spatial organization of human groups, may be expressed as *cohabitation, répartition, réseau des acces, circulation, iconographie.* English equivalents would be functional structure, distribution, transport network, circulation and group policy. This procedure demands a thorough assessment of politics and legislation. Moreover, says Gottmann, it is not enough to have a method, one must always have an objective of research (*un but de recherche*); and this, he concludes, lies in the best political organization in terms of co-operation and justice in the organization of geographic space. This conceptual framework, which I sought to develop and apply in various studies through the thirties, fits closely with current thinking in the United States, but it is surprising that the same conceptual framework can bring forth such differing results, for the so-called regional approach, that is developed by Gottmann, is in disrepute among his American colleagues. It is quite clear that the meaning and implications of the regional concept are in need of reappraisal in the United States.

ANDRÉ CHOLLEY

André Cholley (1886–   ) has long been a leading French geographer and was the Director of the *Institut de Géographie* in Paris from 1944 until his retirement in 1956. The following notes are based on a long personal communication received from M. Cholley on March 8th, 1967, in his eighty-first year.

Born in 1886, Cholley attended school from 1897 to 1903 and then proceeded as a student to the University of Lyon. He took the *licence* in history and geography in 1906 and the *agrégation* in 1910. He started his career as a teacher in the *lycée* of Annecy in 1912, and, after military service throughout the First World War, resumed this profession at the *lycée* in Lyon where he remained until 1923. In that year he took his doctorate in geography with a thesis on a geo-morphological theme on *Les Préalpes de Savoie (Bauges et Genevois): Étude de géographie régionale,* published in 1925. He joined the geo-graphy staff at the University of Lyon as *maître de conférences* in the Faculty of Letters in 1923 and moved to the Sorbonne in Paris in 1927. Here he remained for the rest of his career and succeeded de Martonne as Director of the Institute of Geography in 1944 and in the same year became the President of the Association of French Geographers, a position he still holds. Thus, Cholley chronologically belongs to the second generation of geographers in France, but he entered the profession rather late and reached the peak of his career in the ten years following the end of the Second World War, so that he is more appropriately located with the third generation. His career and views afford a clear insight into the course and trends of geography in France.

Cholley writes that he was trained as a geographer in the *véritable floraison de travaux de géographie régionale* associated with the teaching of Vidal de la Blache, but the master who had the greatest influence in moulding his development (he writes) was Emm. de Martonne. He was one of de Martonne's earliest pupils at Lyon and became his collaborator in later years in Paris. Emm. de Mar-tonne, writes M. Cholley, took regional geography a big step further by revealing *une morphologie rationnelle* in his classic work on physical geography published nearly sixty years ago. He encouraged Cholley to become the successor of Lucien Gallois in Paris and leave his seat in Lyon. Cholley points out that the number of students of geography at Paris before the First World War was only 50 at a maximum. It increased to 100 in 1930 (shortly after his arrival in Paris) and 250 in 1950, immediately after the Second World War. Not only has there been a remarkable post-war increase in numbers of students, but the aims of research have shifted. Geographical research in the thirties was turning in its physical aspects towards more quantitative analysis so that the teaching had to be modified to meet the numbers of students and the advancement of geographical research. This adjustment was begun by de Martonne after the First World War, but it was a slow process.

Cholley's morphological expertise, as applied in the Pre-alps, was transferred to northern France after his move to Paris in 1927. He sought to determine the phases of evolution revealed by the correlation between surfaces of erosion and plains formed during marine transgressions of the same epoch. From many studies in the Paris Basin he discovered a series of peneplained surfaces resulting, at the end of the Miocene, in the establishment of a 'polygenic surface' that affects the whole *ensemble* of the Paris Basin. Following on the tectonic movements of the Pliocene period, the existing surface forms are to be attributed not to such extensive structural and erosional surfaces, but to slopes (*pentes*)—valley slopes, structural talus slopes, that were imprinted by the climatic oscillations of the Quarternary period. This research, continued by his pupils, such as Beaujeu, Tricart, Pinchemel, De Planbol, and M. Birot, was synthesized in a publication in 1946. He is now preparing a regional geography of the Paris Basin.

Teaching in Paris, as elsewhere in France, he continues, involves a double purpose—the training of future secondary school teachers, and the direction of research. This has been rendered difficult by lack of personnel. The institute had 4 professors and 1 assistant in 1956. Today it has 15 professors and 35 assistants. Cholley has been much concerned with geographical education and in 1942 his little book on *Guide de l'Etudiant en Géographie* was published.

A veritable renewal (*renouveau*) of geography in France took place between the wars. Its keynote was the increased emphasis given to the quantitative explanation of phenomena (causes and effects)—a tendency that was found in both the natural and human sciences. 'It demands in morphology, for example, the analysis of deposits and soils, and, for human and regional geography, the use of statistics and the organization of personal enquiries (*enquétes*) in the form of samples (*sondages*).'

Vidal de la Blache planned the new building of the Institute and its organization after its opening in 1923 was the task of de Martonne, though he was seriously hampered by lack of funds. This was remedied after the Second World War, and the process of expansion was the task of M. Cholley. The chief achievement, says M. Cholley, was the organization of a *Centre de Documentation cartographique et géographique* with the financial aid of the National Centre for Scientific Research (C.N.R.S.). During Cholley's directorship, various new publications were begun—*Mémoires et Documents du Centre de Documentation* (1946), *Information Géographique* (1936), *Collection Armand Colin* (originally directed by A. Demangeon) and *Collection Orbis*, a series of volumes on major areas of the world.

Cholley submits some thoughts on current trends in geography in France. There has been much progress in the equipment of Institutes of Geography and research has experienced thereby *un nouvel élan*. Quantitative methods have proliferated in geomorphology and phytogeography, since their establishment in the physical field by de Martonne. Similar developments have taken place in the post-war years in the field of human geography though they are thwarted by the lack of local data. This current trend, however, brings difficulties and dangers. Data, in the field or in the form of statistics (often for administrative units that are far outmoded and that often require adaptation or the pursuit of time-consuming personal enquiries) are difficult to come by. Months, even years, may be required to get them together, even with the help of the *Institut National de la Statistique et des Études Économiques*. It thus becomes difficult to pursue the geographical study of an area in all its aspects according to the *définition vidalienne*. Problems need to be more sharply defined in terms of content and method and practicability. Over the past ten years we have found ourselves confronted with fragmentary studies concerning one of the elements in the structure of an area and its life—urbanism, agrarian structure, demographic movements etc. He concludes in his personal report as follows, and his words are reproduced in their entirety.

> In brief, the idea of regional synthesis disappears before a fragmentation that seems to lead to nothing in the present state of affairs. But the study of particular areas alone, following upon successive reductions of scale, permits us to reach the recognition of continental zones, domains of civilizations, or modes of life, and to reach the reasoned description of the planet, which is the very aim and purpose of geography. The deep analysis of fragments is valid in geography but only on condition that it cedes place, in a second phase, to regional synthesis. This, indeed, was the analytical and synthetic procedure of Blanchard in his work in the Alps. Institutes of Geography must facilitate these goals of analysis and synthesis from a local to a world-wide scale. Regional geography must be re-established in its rights and missions to bring together the materials of a general geography of world-wide applicability.

21

Post-War Trends

GENERAL

The post-war years, wrote M. Sorre in 1957, have brought forth a change in attitude from the tradition of Vidal de la Blache, and new research problems and procedures. There has occurred a 'mutation' in all countries and it is associated with changes in technologies. Objectives and methods of geographical research are being adapted to changed times, but they are rooted in the research procedures of the last generation. Analytical procedures are becoming more pronounced. This trend was already evident in the field of agrarian geography before the war through the stimulus of certain historians but it has spread since the war to the whole field of geography. The reason is undoubtedly the increasing demand for more rigorous scientific procedures and it is also associated with the development and refinement of techniques, both statistical and instrumental. This is part of a general trend in scientific work.

This trend leads to more intimate association with marginal fields of enquiry of both Nature and Man. This is reflected in the extension of the geographical method into the areas of interest of neighbouring disciplines, and fields of enquiry have often been snatched from the geographer though he gave them the first impetus. The geographer must therefore see to it that he develops more assiduously a distinctive contribution to knowledge. Sorre claims that this uniqueness lies in the idea of synthesis, the notion of 'localized ensembles', the notion of interdependence, and of solidarities within such complexes. This goal was the central purpose of the great geographers of the past generation.

Sorre considers that the risks of losing this goal of regional synthesis in the face of analytical objectives is not as great as one would imagine. Among the younger generation, he writes, there remains a central concern with the localization of the phenomenon that are being analysed. Further, one finds increasing cognisance in related disciplines, such as climatology, botany, and sociology, of the concept of the geographic *ensemble*. Spatial phenomena cannot be explained simply by correlation with another set of spatial phenomena, as Vidal de la Blache expressed it. They must be interpreted as a part of a geographical whole (*ensemble*). Herein lies a guarantee for the continued demand and justification for the geographical approach.

GEOMORPHOLOGY

Morphology, writes Mme. J. Beaujeu-Garnier, has undergone quite remarkable changes in the last twenty years. Hypotheses of earlier years have been subjected to quantitative analysis. This has been achieved in the first place by laboratory analysis, whereby processes can be simulated, and hypotheses based on fragmentary observation in the field can be tested. The *Revue de Geomorphologie Dynamique* is almost the official organ of these researches. Using the new techniques of a 'dynamic geomorphology', new problems have been tackled. The laboratory examination of sands, gravels, or clays permits conclusions on modes of transport and ablation, and on the climatic conditions under which the deposits were formed. Modes of disintegration of rocks, both mechanical and chemical, have shifted attention to the effects of climatic conditions on the alterations of limestone and crystalline rocks, and to the effects of cold climates on the development of karst. In this way, a climatic morphology has developed. From recent studies there emerges the general conception of the *dynamisme intégral de la morphologie*. Earth movement is constant and gradual and there is no clear separation between erosion and construction. These processes operate simultaneously. One or other may dominate, or the general evolution may be interrupted, by either tectonic or climatic causes. Cycles initiated at different dates operate simultaneously (polycyclic relief). Processes of desert erosion have been worked out in the field in the Sahara, by jeep, and by plane, and by laboratory simulation.

CLIMATOLOGY

No outstanding work has been undertaken since 1926 says Ch. P. Péguy. In that year there appeared Bénévent's thesis on the climate of the French Alps and de Martonne's important concept of the index of aridity. Since then, climatological work has pursued more specialized lines.

Post-War Trends

The revolution in atmospheric research has led to the development of dynamic meteorology. In France there was a steady reduction in the number of recorded rainfall stations. *The Annales du Bureau central météorologique* listed 930 stations in 1890 as compared with 365 in 1930. 'The meteorologists having resolutely followed the way of physics, have their feet less and less on the earth', and there was little collaboration between the meteorologist and the geographer.

A reaction set in during the thirties and has continued since. De Martonne threw new light on climatological questions by his ideas on *zonalisme*. Attention turned to the relevance of climate to morphogenesis, and today men such as Birot and Tricart speak of 'climatic morphology', in contrast to 'structural morphology'. Similar studies in vegetational ecology have served as a reorientation and stimulus to climatology. The development of hydro-electricity has also stimulated the collection of more stations for meteorological observations.

New concepts and data in meteorology were used in the thirties to explain climatic conditions and this trend has been accentuated since the war. The meteorologists became increasingly concerned with climatic phenomena, as is revealed in their work on agricultural meteorology. Medical meteorology and climatic-therapy have not yet been much developed. Two doctorate theses on regional climatology have been published since the war, one on the Massif Centrale and the other on the basin of Paris.

SOILS

'Like geography, pedology becomes more and more synthetic', writes L. Gachon. This trend has been encouraged in France by the great physical diversity of the country, and consequently by the different combinations of soil affecting processes through time, including the agency of man. The assumption by Russian scholars that climatic determinants are of paramount importance has been rejected, nor is it possible to explain the origins of soils simply in terms of derivation from bed-rock. It has been found that genesis and development of soils in areas of gentle slopes and plains involve the infiltration of waters, solifluction, soil creep, the formation of alluviums and colluviums, capillarity, evaporation, transpiration and the development of eluviated horizons on the surface and illuvial surfaces in depth. Soils, of course, are profoundly modified by man, both by interference with these processes and by changes induced through cultivation of the soil.

Gachon notes that there is, however, a tendency for the scientist to regard soil as so much matter to be tested in the laboratory, but *le sol est vie en même temps que matière.* He suggests that two factors are predominant in the differentiation of the eluviated and illuviated horizons—age and the internal regime of water. For example loessic soils are fertile because they are young, with well proportioned contents of siliceous, clay and lime elements, but they are also differentially affected by climatically induced variations of water action, by the circulation of water itself wherein upward movements dominate in the summer and slow descents in the winter, and by variations of temperature and humidity in both seasons. Hence is derived the fertility of black earths (chernozem) and loess, as well as the increased fertility of soils on granite, as for example in the Central Massif, where on gentle slopes in the mid-altitudes, the soils have been brought into cultivation. Thus the best conditions for soil occur where there is free vertical and horizontal movement of water.

In addition the French pedologists have considered man's transformation of the soil; a study which amounts to a history of conservation and improvement. The remarkable powers of regeneration of which the soil is capable, despite man's maltreatment, has necessitated thinking of soils not in terms of geological periods, but in terms of hundreds of years (as in Flanders and Friesland). However they can, of course, be destroyed much more quickly by man's struggle to feed himself.

BOTANICAL GEOGRAPHY

Flahault (1937) continued in the Mediterranean area the field of enquiry of his nineteenth century forerunners. He mapped vegetation as an outward form and his approach was essentially ecological. It has been continued by Gaussen in his small scale maps of France (*Annales*, 1938) and in his large scale maps (1:200,000) of selected areas.

A second and more recent line of work is the phytosociological approach. This is the minute statistical analysis of groups of species constituting an association which occur at different levels and are mapped on a large scale (e.g. 1:10,000). A current problem is to merge these two modes of approach, the first ecological and broader, the second, statistical and pursued on an extremely minute scale.

A number of exhaustive regional monographs have appeared on these lines of study, including Gaussen's study of the eastern Pyrenees

(1934), and various monographic studies in French North Africa. The latter are brought together in Aubreville's major study of the climate, forests and deserts of tropical Africa (1949).

HYDROLOGY AND OCEANOGRAPHY

M. Pardé, professor of geography at Grenoble, has made the field of hydrology in France peculiarly his own. Advances in such study have been facilitated over the last 25 years by hydrometric and meteorological observations relevant to the explanation of the river regimes.

A geographer, Ch. P. Péguy, has argued for the use of mathematics in hydro-meteorological studies and he employs such methods in his doctorate thesis on the Durance and Ubaye rivers (1947) and in a second work on the introduction of statistical methods in physical geography. This is a highly specialized field that is becoming more closely allied with statistical techniques and closely allied with meteorology as a branch of geophysics.

Twenty years ago the study of oceanography had sunk to a very low level, but in the post-war years, together with coastal morphology and hydrology, this field has experienced a remarkable development, principally in the various departments of the government. A. Guilcher, reporting on this, takes pains to delineate what the geographer can contribute to the field of oceanography, and stresses that its data must be based on observations by organizations on a large scale. The facts and findings of oceanography have geographical repercussions at every turn, but methods of interpretation are mathematical and lie outside the normal competence of the geographer and one cannot escape the conclusion that oceanography is a research field in itself. The professional geographer by and large, is interested in its relevance to man, rather than for its own sake, although the study of geography theoretically defined includes the oceans, their waters and their depths and there is no question that researches on the sea floor and content and movements of oceanic waters will always be acceptable to geographers. The theoretical field is just too big, and substantial sections must be conceded to specialists.

Specialist courses on the ocean are given in some departments of geography in France, such as A. Guilcher's courses at the University of Nancy, and geographers have also been instrumental in the past in the development of oceanography, but that time is passed. The field now ranks as one of the major branches of geophysics.

France

HUMAN GEOGRAPHY:
THE AGRARIAN LANDSCAPE

In the general field of human geography pride of place must be given to the work of the late Maximilien Sorre that has been already noted in Ch. 19.

The agrarian landscapes have been investigated in a remarkable work by Et. Juillard on Basse-Alsace. He points out that the country-side, *les campagnes*, ranks high in geographical research in France. Between the wars attention was directed to two main questions, the technical and economic problems of agricultural activities and problems of the organization of rural space, and these trends have continued since the war. However new ideas are now being pursued and more detailed analysis of particular situations reveals the complexity of the rural landscape and its moulding forces. Documentary sources, cadastral maps, and serial photographs have facilitated such advances and they have also been aided by developments in phytosociology, pedology, toponomy, ethnography and demography. Historians are beginning to reconstruct past landscapes, and are being called 'geo-historians'.

There are three trends in the post-war work of geographers on agrarian landscapes. One may concentrate on a past period, as Dion did, on the Middle Ages, or Champier on pre-history. Others may concentrate on the technics of cultivation and peasant mentality (Faucher and Gachon). Still others have used statistical methods. Several of these trends may be summarized. Demangeon's pre-war survey of farmsteads is now being advanced by documentary study of the development of building structures. The diversity of forms in agrarian morphogenesis has been revealed, but we are still far from reaching comparative conclusions, and although the treatment of the countryside as a whole has been aided, there is still much to do. The landscape cannot be interpreted simply in the light of physical environment and the human elements. 'Man utilises the physical milieu through the intermediary of a particular civilization' (Gourou). If one civilization replaces another in the same physical context, the human geography will be different. The word civilization (or culture) here means the techniques of using land and organizing space. This idea has directed recent research and is evident in the approach of D. Faucher in his agrarian geography before the war and in his later book on the 'peasant and the machine' written since the war (1954). It is reflected also in recent regional monographs which include the works of R. Dion on viticulture in France (1959), P.

Philiponneau on the rural outskirts of the Paris region (1955), and Et. Juillard on Alsace (1953).

From these researches two general ideas have developed. First, the physical elements of the milieu are transformed by successive human groups; savannas and prairies are in large part man-induced; Gradmann's famous steppenheide vegetation is probably a formation induced through human interference; the destruction of soils by human misuse has necessitated changes in economy or the abandonment of villages. Second, although field systems used to be regarded as fixed and derived from obscure origins in remote antiquity, it is now considered to be more realistic to regard field patterns as the changing expression of human occupance. One investigates open or closed fields, dispersed or clustered farmsteads and methods of communal usage (Bloch and Dion). These are interpreted in terms of changes of landownership, techniques of cultivation, and the succession of agrarian systems.

POPULATION AND PRODUCTION

The geography of population seeks to establish the localization of the characteristics of human populations. It has its modern roots in studies by demographers, such as A. Sauvy, A. Landry and others. Among geographical works with world-wide reference attention is drawn especially to the conceptual approach of P. George and the substantive studies by J. Beaujeu-Garnier (in two volumes). One should also make reference to *in situ* demographic trends and migratory movements that have been examined in particular areas of France. Particularly important in this connection are the researches of G. Veyret-Verner and his wife and their school (Grenoble) in the French Alps. Studies of the same kind have been undertaken in the Central Massif.

The geography of urbanism has been led by G. Chabot in recent years, as evidenced most recently by his work (in collaboration with J. Beaujeu-Garnier) on *Traité de Géographie Urbaine* (1963, 493 pp.). Its foundations were laid in Blanchard's classic study of Grenoble in 1910. In addition to many contributions to the geographic study of particular cities, such as Paris, St. Etienne, and Toulouse, special attention is drawn to comprehensive works on the geography of urbanism by P. George and G. Chabot, and to J. Labasse's remarkable investigation of the role of Lyon in the economic growth of its surrounding area. More recently certain scholars have turned to the hierarchy of urban centres in France (Pinchemel, Coppolani,

Rochefort, and Dugrand) and to the need for further investigation of the impact of the town on the life and organization of its dependent area. Gravier's notion of 'Paris and the French desert' has made a big impact in understanding the current problems of France and draws attention to such problems of urban distribution and service in the countryside.

The geography of industry in France had its first great contribution in the work of P. de Rousiers. This was in five volumes and appeared in 1927–8, however, much of the geographical work on this aspect is to be found in the regional monographs. Special note may be made of later works by J. Chardonnet on great industrial complexes (1962–5) and R. Gendarme's economic study of the industrialized area of northern France (1954). A number of important geographical works refer to a particular activity in France or to a systematic analysis of the localization of certain industries in general, and a series of books is devoted to the latter theme. I will mention P. George's study of the geography of energy (1950) and A. Allix's geography of textiles (1957). Geographical expertise is being applied to practical problems as illustrated by J. F. Gravier's best seller on *Paris and the French desert* (1947) and *Decentralization and Technical Progress* (1954).

The geography of circulation has long been a special concern of French geographers. A definitive work was R. Capot-Rey's *Géographie de la circulation sur les continents* in 1946, and the field was even more adequately formulated by M. Sorre in the portion of his massive work on the foundations of human geography which deals with man's conquest of space (Vol. 2, 1948). A number of studies deal with the role of the route and with the spatial impact of points such as great railroad depots (R. Clozier, *La Gare du Nord*, 1940). A three-volume work on *Géographie de la Circulation* is now published.

THE REGIONAL MONOGRAPH

This approach, wrote, R. Musset in 1938, in the Geographische Zeitschrift, means the mastery of many kinds of processes in order to understand the unity of the whole that might well exceed the capacity of one man and since each country has its unique features, this involves a method of portrayal, a charted course, a plan of salient questions, all of which will vary with the approach and especially with the area. Such was the traditional objective, but in the last twenty or thirty years there have been further changes in the content, purpose and method of these studies.

Post-War Trends

The post-war monograph, and also the monographs of the twenties and thirties, such as by Deffontaines and Dion, reveal a shift away from a so-called 'complete study', that supposedly embraced all geographical aspects, to studies in which one point of view, one objective, and the series of questions arising from it give focus, direction and plan to the investigation. The physiognomy of the area is thus only partially portrayed. What then is the objective?

It seems that the development of general geography is leading to the amplification of findings (i.e. systematic) to a particular area, a procedure that is not without damage to the autonomy of regional geography in the accepted traditional sense. (R. Musset)

There are however, a number of post-war monographs that have attempted the traditional 'regional synthesis'. An example is Gautier's study of central Brittany (1947). More general works include R. Blanchard's monumental eleven volumes on the French Alps (1937–54) and Le Lannou's two volumes on Brittany (1950–2).

Questions to receive special attention include those of morphology. J. Beaujeu-Garnier (1951) revealed that the genesis of the land-forms of the Morvan are far more complicated than was postulated by Davis over fifty years ago; Péguy has used statistical methods to analyse the process of fluvial erosion in the Haute Durance (1947); Tricart has dealt with the morphological development of part of the Paris Basin (1952) and Pinchemel with the relations between the Paris Basin and the southeast of England (1952). More recent monographs cover Madagascar, Galicia, and the Andes of central Chile.

Questions of human geography to receive special attention are those of rural organization and agrarian life. Two major studies alone will be cited, M. Derruau on *La Grande Limagne auvergnate et bourbonnaise* (1949) and Et. Juillard on *Basse Alsace: La Vie rurale dans la plaine de Basse—Alsace: Essai de Géographie Sociale* (1952). Economic emphasis is found in Veyret-Vernier's work on industry in the French Alps (1948) and A. Journaux's on electricity in Lower Normandy (1955). The development of house types in the northern French Alps was the subject of J. Robert's work in 1939. R. Dugrand writes on the urban centres of Bas-Languedoc (1963) and M. Rochefort on the urban geography of Alsace (1960).

There is a basic problem of relevance in the geographical study of a small area. The individual facts—fields, plant pots or social attitudes—must be examined in terms of their spatial patterns and associations, that is, their range and localization. Otherwise, one gets bogged down in a welter of detail. If no such relationship or pattern is found to exist, they should be jettisoned from the geographical framework. To attain this end, a definite central theme and thus a rigorous selection is found in recent studies, such as that of Max Derruau on the *Grande Limagne* (1949), around and north of Clermont Ferrand. Its keynotes are the types of terrain and the patterns of settlement imposed upon them and the way in which, historically, these contrasted patterns of human occupance are imposed upon the variety of local terrains.

R. Dion writes in the *Annales* as follows in a review of Max Derruau's monograph: 'Anyone who wishes to have a clear idea of the most recent and the most decisive accomplishments in France of geographical investigation should read the book of Max Derruau; this is at present one of those which ought to be recommended as of primary importance.' The purpose of this study, he continues in the same review, illustrated by a remarkable map of the existing fields patterns, is 'to explain the human imprint on the surface of the land'. It was generally assumed down to 1930, writes Dion, that to pass from one area of clustered villages to one of dispersed farmsteads meant *une limite naturelle*. This simple interpretation was laid low by Marc Bloch's remarkable work on the history of rural France in 1931, which demonstrated that landscapes depend on changing agrarian structures, that are revealed in the 'parcels' observable on cadastral maps and serial photos. The two types of distribution, agrarian structures and natural divisions, are fundamentally different. M. Derruau's map, noted above, reveals this fact, and much of his work is devoted to explaining it.

In recent years a number of large volumes on the geography of particular areas of the world have been, and are being, published. Special mention is made here, as an example, of the eight volumes in a series on France called *France de Demain* that have appeared since 1960 and the *Orbis* Series which includes several major volumes on Europe (Central Europe, Northern and Northwestern Europe, Mediterranean Lands), with other volumes in preparation.

AN APPRAISAL OF CURRENT TRENDS

In conclusion, the following paragraphs are an appraisal of post-war trends in France by one of its younger generation, Philippe Pinchemel, until recently Professor of Geography at the University of Lille, now professor at the Sorbonne. Pinchemel is particularly suited for such an appraisal, since his doctorate work was in geomorphology, but in recent years he has turned to questions of human geography, most notably to a functional study of urban settlements in France, for which work he has employed statistical methods. The present writer's assessment in the proceeding chapters has been written quite independently of Professor Pinchemel's and it is therefore of special interest to note points of similarity and divergence in the two assessments. Professor Pinchemel writes as follows, in free translation:

The first trait to be emphasized is the extraordinary fecundity of French geography. This is evidenced by many new professional chairs and teachers in geography, the increased number of publications and theses, foreign missions and communications to international congresses. This expansion rests on the solid base, ancient and deep rooted, of teaching at all levels. From the age of 5 to 17/18 geography is the only human science in the school syllabus. Associated with history it opens to the pupil the doors of the contemporary world. The appointment of professors of geography is to meet the essential aim of higher education. French geographers devote most of their time to works of synthesis and to *mises au point*, works that bring to the public solid manuals of general and regional documentation.

Parallel to this dynamic growth, however, one does not find a unanimous feeling of security about the future of the discipline. Words of *malaise de crise* recur periodically from the pens of certain geographers. To this attitude, there are to be added criticisms addressed less to geography than to certain geographers, who are reproached by the specialists for intervening in *multiple domaines* without having the competence or the expertise necessary to cope with them. It is evident among other sciences or disciplines that the position of geography is neither proportioned to its university status, nor assured of its objects and methods. Its status is thus a matter of controversy.

Another feature of geography in France must be emphasized. Until 1945–50 geography lived in splendid isolation, assured of its superiority in virtue of the brilliant succession of geographers since the days of Vidal de la Blache and his leading students—Emm. de Martonne, A. Demangeon and R. Blanchard. Thus, French geographers did not keep abreast of trends of thought in other countries.

This double contradiction—between teaching and research, growth and uncertainty—is the reflection of its expansion in all directions with the inevitable consequence of specialization. The enlargement of geographical curiosity is unquestionable. Geographers have described more and more numerous phenomena in all parts of the world. Banks, investments, quaternary morphology, biogeography, pedology, glaciology and karstology, are among the relatively new themes taken up by geographers. The traditional theses have become more precise in their objectives and methods. In agrarian geography, for example, the analysis of rural landscapes (field systems, land-ownership, enclosures) has been the subject of numerous investigations. In urban geography, parallel to urban monographs, geographers have published studies of regional or national networks.

French geographers have worked in recent years in Greenland, Spitzbergen, Japan, Afghanistan, Turkey, India, Australia and in various states of Africa and America. Geographical expeditions have been carried out in high latitudes, in karst regions and in the Near and Far East.

Recent researches have been oriented less towards description and descriptive analysis in a regional presentation, and more towards explanation with a genetic perspective. It is the origin of *bocage*, for example, rather than its typology that now draws attention. This tendency in human geography has been favoured by the historical formation acquired in the Faculté and the close links between historians and geographers.

In physical geography, the study of relief and climate has drawn many researches desirous of understanding and explaining the process and systems of erosion; and of reconstituting the stages of morphogenetic evolution. Superficial deposits are as important, if not more so, than forms of relief. Similar interests have been manifested in climatology, emphasizing the understanding of the atmospheric circulation, rather than the minute analysis of elements of climate, local or regional. The same development is apparent in hydrology, biogeography and pedology. All these researches and their advanced techniques involve high specialization. In physical geography, the analysis of deposits, gravels, sands and clays has occasioned the establishment of morphoscopic and granulometric laboratories. At Nancy, Paris and Caen centres of this kind have been established. At Grenoble a climatological laboratory is under construction and centres for pedological researches are envisaged. Geographers are also involved in research on the process of weathering.

In human geography, emphasis has been placed on economic factors of production and consumption, on economic systems and regimes, on zones of influence, economic regionalization, the analysis of traffic flows, far more so than on the relations of these with the natural milieu.

Post-War Trends

The main consequence of this orientation and specialization has been a weakening of the links between physical and human geography. Each of the foregoing fields has progressively advanced towards autonomy, and is less concerned with providing to the other branches explanations or correlations, withdrawing from geographical researches that which has seemed to the present to be an element of unity.

Thus, geographers, with increasing specialization, have moved more and more in two distinct universes, physical geographers in the universe of the natural and physico-chemical sciences, and human geographers in the universe of demography, economics and sociology. The majority are little concerned with questions of precise definitions of their field. They regard geography as less an object than a mode of approach (*moins un objet qu'un etat d'esprit*, as H. Baulig put it some years ago).

To express this trend, one should speak of 'geographical sciences', rather than one 'geographical science' such as is held by a minority. This significant minority of geographers defends vigorously the view of a single geographical science of the organization of space or landscapes. They follow the tradition set by Vidal de la Blache and Max Sorre. The specialized geographers of agrarian geography are among these partisans of a geography definable in terms of single objectives.

Regional geography suffers from this (current) specialization. It is rarely considered as a framework for true scientific research. It persists in works of popularization for 'the intelligent layman'. Nevertheless, regional geography has not lost as much as one might imagine. There are those who remain convinced of the originality of a synthetic approach to regional problems. This approach in regional geography benefits from competition with other fields that are concerned with the ecological aspect, such as economics and sociology, since it has had to redefine with precision its objectives and its field of study.

To the conception of a regional geography of an encyclopedic order, bestowing attention on all that is present within the limits of the area of study (regional) there has developed a more limited objective. Regional geography is now more concerned to describe and explain the complexity of the organization of space. Regional geography is helped towards this goal, first, by the use of detailed maps, and above all by aerial photographs; and, second, by the growth of attention to political and economic planning.

The applications of regional geography are equally, if not more important, than those of general geography. Whatever the type of geography in question, its methods become more and more scientific, both in physical and in human geography, in which statistical methods are making a slow, prudent and irreversible entry.

The view of French geography in 1964 is far removed from that of thirty years ago. Its reputation rested then on a small number of professors with similar interests publishing their work as theses, the *Géographie Universelle*, and the *Atlas de France*. Paris and Grenoble were the two poles of French geography. Today Paris still has a predominant influence, on account of personalities, equipment and powers of command and animation, but geography is undoubtedly much more decentralized today. Provincial institutes have their own researchers, specialized laboratories and centres, and their own publications. Five new provincial periodicals of regional geography have appeared since the war.

Further References

This chapter relies heavily on *La Géographie Française au Milieu du XXe Siècle*, L'Information Geographique, Paris, 1957. This contains a series of articles by distinguished authorities covering each of the special fields of geography. Professor Philippe Pinchemel, has kindly written an appraisal of the situation as of today. Brief, but very much to the point in terms of the purpose and trends of geography in the light of *la tradition vidalienne*, is G. Chabot, 'Les Conceptions Françaises de la Science Géographique, *Norsk Geografisk Tidskrift*, 12 (1950), pp. 309–21.

22

Professional Growth

The growth of geography over the past hundred years, and in particular the French and German contribution to it, is revealed by its place in the universities, where there has been a great growth in numbers of both professors and students throughout the world; and by the organization of national and international bodies to promote it. The university record in France and Germany has been covered in these pages and several enquiries have been published[1], the latest in German by W. Hartke in 1960[2]. The growth of societies has also been examined.[3] The history of the International Geographical Union has appeared in print, though in a source of very limited circulation.[4] We shall comment in this chapter on the second and third matters, and then draw some general conclusions.

The first society to be founded was the *Societé de Géographie de Paris* in 1821. It was followed by the *Gesellschaft für Erdkunde zu*

[1] W. L. G. Joerg, 'Recent Geographical Work in Europe', *Geog. Rev.*, 12 (1922), pp. 431–84. See also J. S. Keltie, *The Position of Geography in British Universities*, Am. Geog. Soc. Res. Series, No. 4, 1921, and Emm. de Martonne, *Geography in France*, ibid. No. 4a, 1924. See also Wright and Platt, *Aids to Geographical Research*, New York. Am., Geog. Soc., 1947, pp. 106–8, and the relevant chapters in G. Taylor (ed)., *Geography in the Twentieth Century*, London, 1951, especially those by G. Tatham and R. J. Harrison Church.

[2] W. Hartke, *Denkschrift zur Lage der Geographie*, Wiesbaden, 1960, *Deutsche Forschungsgemeinschaft*.

[3] On the growth of societies and journals see the following—C. D. Harris and J. Fellman, *Union List of Geographical Serials*, second ed. 1950, Univ. of Chicago, Dept. of Geography, Research Papers, No. 10; J. K. Wright 'The Field of the Geographical Society' in G. Taylor (ed.), *Geography in the Twentieth Century*, pp. 543–65; T. W. Freeman, *One Hundred Years'of Geography*, Ch. 3, pp. 49–68; J. K. Wright, *Geography in the Making*, 1851–1951, New York, 1952; J. K. Wright and E. T. Platt, *Aids to Geographical Research*, second ed., 1947, New York. Am. Geog. Soc.; and H. R. Mill, *The Record of the Royal Geographical Society*, 1830–1930, London, 1930.

[4] This is based on two articles that have appeared in the proceedings of the International Geographical Union translated and published in *Orbis Geographicus*, 1960, pp. 61–97.

Berlin, founded by Carl Ritter in 1828. Two years later the Royal Geographical Society in London came into being (1830). Others were added in quick succession: the Geographical and Statistical Society of Mexico in 1833, the Geographical and Statistical Society of Frankfurt in 1836, the Geographical Society of Brazil in 1838, the Imperial Russian Geographical Society in 1845, and the American Geographical Society in New York in 1852, with the name American Geographical and Statistical Society. In 1866 there were eighteen geographical societies throughout the world and eleven of them in Europe. It is characteristic that these societies were primarily associated with the great phase of continental and oceanic exploration of the century and also with the collection of masses of statistical material of all kinds.

Since 1866 the number of societies has increased rapidly with the great growth of geography as a field of scientific study with university status. In 1901 there were eighty-nine geographical societies in Europe with more than 60,000 members, six in Asia, eight in North America, five in South America, three in Africa and four in Australia, a total of 115. There were more than 150 journals. In 1930 there were 137 societies throughout the world, and their periodicals were published in some forty different languages. In 1960 geographical periodicals appeared in fifty-one countries, twelve of which published more than ten serials each. Germany led the field with fifty periodicals, followed by the Soviet Union with forty-three, the United States with thirty-seven, Great Britain with twenty-seven, and France with twenty-three. At that time there were 110 periodicals in English, compared with fifty-nine in German and thirty-five in Russian.

It is important to follow the history of the international congresses since it clearly reveals the makers of geography at work and the ways in which they have conceived and organized their discipline over a period of nearly one hundred years. There have been twenty international congresses, the first in Belgium in 1871, the last in Delhi in December of 1968.

The first congress was a convention called in Belgium for all those specialists interested in *la science de la terre* in order to honour the cartographers Mercator and Ortelius. Among the participants were August Petermann, the German cartographer, Armand de Quatrefages, the French ethnographer, Ferdinand de Lesseps of Suez Canal fame, Élisée Reclus, one of the few professional geographers of that time, and Vivien de St. Martin, the French cartographer. The programme included 26 questions of geography, 22 of cosmography, 36 of navigation, meteorology, commerce and statistics, and 3 of ethnography. Questions of surveying and map projections, the orthography of place names, and polar exploration dominated the geographical section.

The second congress was held in Paris in 1875. Its work fell into the following sections: mathematical geography, hydrography, physical geography, historical geography, economic geography, education and exploration. The definition of geography approved by the assembly runs as follows:

> The objective of geography is two-fold: first, knowledge of the natural configuration of the earth's surface considered as an end in itself (*géographie physique*); second, the study of the relationship of the earth with its inhabitants (*géographie politique, ethnographique, et économique*).

The participants included Ferdinand von Richthofen (President of the *Gesellschaft für Erdkunde zu Berlin*), Julius Perthes and August Petermann from Germany, Sir Henry Rawlinson (President of the Royal Geographical Society of London); and Ferdinand de Lesseps, Élie de Beaumont, and Vivien de St. Martin from France.

The third congress took place in Venice in 1881. It conducted its business (the reading of papers) in eight sections that formed a basic framework for several of the following congresses. These sections were entitled: mathematical geography, hydrography, physical geography, anthropological geography, historical geography, economic geography, education and exploration.

The fourth congress was again held in Paris in 1889, the year of the *Exposition Universelle*. Its chairman was Ferdinand de Lesseps, president of the Geographical Society of Paris. The topics were divided into seven groups: mathematical geography, embracing geodosy, hydrography, topography, and cartography; physical geography, including meteorology and climatology, geology, botanical and zoological geography as well as medical geography; economic geography, embracing commercial and statistical geography; historical geography; education; voyages of exploration; ethnographic and anthropological geography.

Berne was the site of the fifth congress in 1891. It was here that one of the greatest enterprises of modern cartography was launched. Albrecht Penck proposed a world map on a standard scale of 1:1,000,000. Present were Albrecht Penck, Eduard Bruckner, Joseph Partsch, and Thomas Holdrich.

The sixth congress was held in London in 1895, under the chairmanship of Sir Clements Markham, president of the Royal Geographical Society. We now find eight sections: mathematical geography (including geodosy); physical geography (including oceano-graphy, climatology, and geographical distributions); cartography and topography; exploration; descriptive geography (the first appearance) and orthography of place names; history of geography; applied geography (especially history, commerce, and colonization) (another first); and education. Special attention was given to the teaching of geography in schools and universities, for these were the years when geography was seeking to overcome a general reluctance among scholars to recognize it as *une science autonome*. Commissions were appointed to inquire into the publication of a bibliography of geographical works and the promotion of polar exploration.

The seventh congress, with 100 members present, was held at Berlin in 1899, under the direction of Ferdinand von Richthofen. Scientific work assumed a more important part in its discussions and among the contributors were Hugh Robert Mill, J. Scott Keltie, Friedrich Ratzel, Paul Vidal de la Blache, and Erich von Drygalski.

The eight congress, the first held outside western Europe, took place in the United States. This was a peripatetic congress for it divided its time between Washington, Philadelphia, New York, Chicago, and St. Louis. Its president was Robert F. Peary, the Arctic explorer. Penck's million map went a big step further by an agreement to delegate the task of preparing the sheets of South America to the United States—the work was subsequently taken over by the American Geographical Society, largely through the direction of Isaiah Bowman. A commission was formed to study the question of statistical materials in countries without a census.

The ninth congress was held in Geneva in 1908, with the president of the Geographical Society of Geneva presiding. There were four-teen sections: mathematical geography and cartography; general physical geography; vulcanology and seismology; glaciology; hydro-graphy (potamology and limnology); oceanography; meteorology, climatology, and terrestrial magnetism; biological geography (botanical geography and zoogeography); anthropogeography and ethnography; economic and social geography; exploration; teaching of geography; historical geography; and rules and nomenclature. Major resolutions at the congress included the encouragement of polar exploration, glaciology, oceanography, and topographic surveys. The international map was further discussed and a project was submitted to the congress by Penck (Germany) and Gannett (United States), and approved unanimously by the assembly.

This congress also adopted a motion on the field of geography which ran as follows:

Professional Growth

Geography as a branch of instruction (*enseignement*) has as its object the description of the surface of the earth, considered in its various elements, physical and living (*physiques et vivants*), whose combination and interdependence determine the existing physiognomy of the globe. This teaching in primary and secondary education should be based on the reading of maps and should pursue above all a synthetic method. In the portrayal of the different parts of the globe, geography should make clear the relation between the inorganic world and living things, and more particularly between the surface of the earth and man.

The tenth of these early congresses was held in Rome in 1913. It cut down its activities to eight sections: mathematical geography; physical geography; biogeography (the first time this term appears in the professional vocabulary); anthropogeography and ethnography (Ratzel's term has taken root but at the time a strong link persisted with the work of ethnographers); economic geography; chorography (this appears for the first time as the theme of a section and, though rooted in German work, has not appeared since); historical geography and the history of geography; and methodology and teaching (the first time the latter makes an appearance). The commissions appointed at Geneva reported to the congress. The commission on the International Map decided to call a conference in Paris in December, 1913. The commission on relief forms assumed a new importance, and encouragement was given to the promotion of oceanographic research in the North Atlantic. A move to organize an international union of geographical societies was delayed by the outbreak of war in the fall of 1914.

The international organization was re-established at Brussels in 1922 as the International Geographical Union at an assembly of the *Conseil International de Recherche*. This Union was to be a professional organization with three objectives—to promote the study of geographical problems; to initiate and co-ordinate research requiring international co-operation, and to provide for their discussion and publication; and to organize international congresses and research commissions.

The first meeting of the new Union, and the eleventh international congress, was held at Cairo in 1925. Work was divided into six sections: mathematical geography; geodosy and cartography; physical geography; biological and human geography; anthropology and ethnography; history of geography and historical geography. While previous congresses had been called and organized by the geographical societies, the new International Geographical Union

271

was based on national committees in each member country. The Cairo congress was in effect a transition from the old to the new regime. The Assembly resolved that its work raised two priorities—an international geographical bibliography and a speedy continuation of the International Map. A commission was approved for the study of the rural habitat, and work continued on the publication of an atlas of forms of relief (initiated at Geneva).

The meeting in Cambridge in 1928 had Sir Charles Close as Chairman. There were six sections: mathematical, physical, biological, human, historical, and regional geography. Regional geography (in place of 'chorography') appeared for the first time as well as human geography. Commissions made reports on the rural habitat, the International Map, and Pliocene and Pleistocene terraces. Several other commissions were established to study climatic variations, population, the mapping of the Roman Empire on a scale of 1:1,000,000, and to make an inventory of ancient maps.

The congress at Paris in 1931 was outstanding since it became a model for subsequent congresses. Preparation was in the hands of the secretary of the national committee, Emmanuel de Martonne. The work was organized in six sections: topography and cartography, physical geography, biogeography, human geography, historical geography, and bibliography and teaching.

The section on physical geography turned to the study of local climates, glacial erosion, tertiary peneplain levels in Europe, river capture, karst erosion, erosion in tropical lands, and continental dunes. The section on human geography considered the localization of industry, urban agglomerations, the distribution of settlements in tropical lands, migrations in countries with unreliable rainfall, methods of representing population on maps. The section on historical geography (still unsatisfactorily defined) turned to the origins of modern cartography—mainly, in other words, to the history of maps. Particular attention was given to the excursions in different parts of France. The number of participants reached 900, with 350 from France, and 42 countries were represented. The number of member countries reached 27. Several new commissions were approved by the Assembly—aerial photography, overpopulation, and the cartographic representation of erosion surfaces—and the existing commissions were reapproved. These nine commissions continued to work actively until the next congress at Warsaw in 1934.

The Warsaw congress was run very much on the lines of the Paris congress. Its sections were: cartography, physical geography, human geography, prehistoric and historical geography, geographical landscape, and didactics and methodology. The congress at Amsterdam in 1938 also continued the organization initiated at Paris.

The Second World War interrupted the meetings of the Union. After the war the executive committee refilled its ranks with de Martonne as president, and the next congress was called to Lisbon. There were seven sections. Two of the commissions were dropped (climatic variations and aerial photography) but new commissions were approved for aerial photography, regional planning, soil erosion, and periglacial morphology. De Martonne was elected Honorary Life President in recognition of his great service to the Union.

This congress at Lisbon, the first after the war, marked the end of a period in the history of the Union. In the interwar years the European influence had been predominant, and again and again one detects the influence of de Martonne in its scientific work. Now the centre of gravity shifted towards the United States.

The congress at Washington divided its work among twelve sections: biogeography, cartography, climatology, demography, morphology, hydrography, historical and political geography, regional geography, resources, teaching, commerce and transport, rural and urban settlement. The assembly approved the continuance of commissions on medical geography, periglacial morphology, ancient maps, and world land use survey. New commissions were elected on arid zones, karstic phenomena, development of slopes, levels of erosion around the Atlantic, coastal sedimentation, the teaching of geography in schools, and the methods of classification of geographical literature.

The next congress was held in Rio de Janeiro in 1956. The papers were grouped into twelve sections: cartography and photogrammetry, geomorphology (most active section, especially in respect of quantitative methods), climatology, biogeography, hydrography, human geography, population and settlement, agrarian geography, medical geography, economic geography, historical and political geography, and regional geography. The existing commissions were renewed and to them were added new commissions to study national atlases, applied geomorphology, a map of world population, and the humid tropics. A new departure was the establishment of regional congresses between the main meetings. There have been such meetings at Kampala, Uganda (1955), Japan (1957), Malaya (1962), and Mexico City (1966). Commissions have also had their conferences of a week or more.

There remain for consideration the meetings at Stockholm in 1960 and in London in 1964. Here we find the Union truly international, and its activities reflect the great range and high quality of researches by hundreds of scholars from all over the world. At Stockholm Carl Troll was elected president for the next four years and he presided over the meetings in London in 1964. It was decided in London to hold the next congress in New Delhi in 1968.

Registration at Stockholm had reached 1,670 and 64 countries were represented (attendance at Rio was 1,000 and Washington 1,300). Registration at the London meeting just exceeded 2,000, drawn from 67 countries (700 from the United Kingdom, 307 from the United States).

At the Stockholm meetings there were ten sections: geographical cartography and photogeography; polar and sub-polar geography; climatology and hydrography; oceanography and glaciology; geomorphology; biogeography; human geography; economic geography; methodology and bibliography; applied geography. Fifteen commissions reported on their work and all were continued for the next four years plus new ones on cartography and economic regionalization. The symposia, which lasted from three to ten days, were concerned with the following themes: the historical development and present character of agrarian geography, coastal morphology, land use, settlement, industrial centres, irrigation, the coast and the archipelago (held at Helsinki), rural settlement in Norway, nature and culture in the western fjords of western Norway, nature and economy of northern Norway, fluvial morphology, expansion and retreat of settled areas, high mountain regions, the Finnish lake plateau, the *zentralort* and its functions in urban geography.

Nearly 1,000 papers are summarized in the volume of abstracts for the London meetings. There were 40 excursions before and after the congress and 11 exhibits in London. At London, five commissions were terminated (ancient maps, library classification, karst phenomena, cartography, and erosion surfaces around the Atlantic), and five new commissions were approved—agricultural typology, aerial photographs, applied geography, international hydrological decade and quantitative methods, and world population. Commissions continued on world land use survey (1949), arid zones (1952), teaching (1952), national atlases (1956), humid tropics (1956), medical geography (1949), periglacial morphology (1949), development of slopes (1952), applied geomorphology (1956) and economic regionalization (1960).

Ten symposia were held before the London congress at different bases in Britain and four after the congress. They covered a wide field of topics—geomorphology (3), urban geography (2), rural settlement (2), cartography (2), geography in the tropics, colonization, agricultural geography, teaching, political geography, and industrial geography. There were nine sections: population and settlement; economic geography; climatology and hydrography; oceanography and glaciology; biogeography; geomorphology; historical geography; applied geography; regional geography; cartography.

A special note should be made of the bibliographical work of the I.G.U. An international geographical bibliography, beginning in 1891, was the work of the *Annales de Géographie*. The first volume under the auspices of the I.G.U. was published in 1949 and now runs into over 70 volumes. The first volume of the international carto-graphic bibliography appeared in 1946–7. It likewise began in France and now runs into 18 (1967) volumes. The International Cartography Association was affiliated with the I.G.U. in London in 1964.

From these hundred years of history of International Congresses, we make the following comments.

The fifty years from 1871 were primarily dominated by the geographical societies of Europe. They were run primarily by 'edu-cated laymen', as well as surveyors and cartographers who found the society and the international congresses organized by them the best forum for discussion of their work. In the second fifty years, since the First World War, the congresses have been run by professional geographers and organized strictly as a professional organization. The views of the minority of professional geographers in the pre-war congresses gradually made themselves felt. In the inter-war years, their views became paramount and were crystallized and forwarded by the influence particularly of French geographers, notably by de Martonne. Since the Second World War, as reflected in the 1964 meetings in London (which were inaugurated by the Queen in the Royal Albert Hall), geography has grown greatly in strength but there has been a trend for its researches to be canalized in particular fields.

A second observation is the change in the scope and depth of the field. In the first fifty years, as the sections indicate, the field of enquiry was vast, but the small band of professional geographers, notably those from central Europe, sought more precisely to define it. The fact is that this was an international forum for researches in a diversity of fields to whom alternative outlets at that time were not available. Since World War II international conferences in other fields have been established, such as geodosy, photogrammetry, cartography, history of science, soil science, population, etc. There has been also a distinct trend for fields of research to shift and crystallize around certain questions. It should be noted that scholars in certain fields still find their most sympathetic milieu in geography. An example is photogrammetry. One also notes the long persistence *ab initio*, and with increasing diligence and clearly defined purpose, of 'physical geography', and especially geomorphology, as it has come to be called, in the last fifty years. A research commission on pliocene and pleistocene terraces persisted for many years between the wars and highly specialized commissions in geomorphology are today very active.

A third observation is the increasing emphasis on the development of techniques and bibliographies that are germane to geographical science. This is very evident in the commissions approved in 1964—quantitative methods, photo interpretation, agricultural typology, and the world land use survey. Research on settlement— the spatial arrangement of buildings, farms and fields—has long been a central concern of geographical research and was especially prominent between the wars in France and Germany.

A fourth observation of the general trend of research interest deserves the strongest emphasis. Physical geography, in its various aspects, has remained a basic concern of geographers throughout the world for the whole period of one hundred years. The content and purpose of physical geography was made clear by de Martonne in his pioneer volume in the 1900's. At the congresses between the wars and especially during the past twenty years a large and increasing proportion of all the papers offered are in the fields of physical geography, notably geomorphology, climatology, oceanography and hydrology. This intense activity is also evident in the research, stretching over many years, of several of the commissions. This predominance and high degree of specialization follows from an uninterrupted European tradition established by men such as Penck, Richthofen, de Martonne and Baulig. Hettner also began as a geomorphologist. These men had their American contemporaries in W. M. Davis, J. W. Powell, G. K. Gilbert, and D. W. Johnson. During the past thirty years, however, these fields in America have been relinquished, with the conspicuous exception of the Berkeley stronghold. American geographers, in consequence, express surprise, a little impatience, and much concern at the dedication of their colleagues to problems of physical geography.

It is also to be noted that the human aspects of geography have assumed increasing prominence in the last fifty years. There is a proliferation of papers offered on these aspects, but there is a dearth of stimulating research or exciting discussion. Sections earlier entitled 'ethnographic' or 'anthropological' have gone, but the separation of 'historical geography' from 'history of geography', with distinctive fields of interest, has only recently emerged, and their growth seems to have been checked, in part at least, by confusion with the history of cartography and the cataloguing of ancient maps. At no congress has there been a section devoted to a major world area, so much so that a contributor often has difficulty in deciding on the appropriate section in which to present his paper, while the section on 'regional geography' has papers that could just as well be presented elsewhere. It is, therefore, of special interest to note (in commissions and symposia) new ideas that are being explored. Such are the historical development of settlement, the areal impact of urbanism (though many are seduced by classification as an end in itself), and the application of new techniques, especially photographic interpretation, quantitative techniques and agricultural typology.

CONCLUSION

The main lessons we learn from the makers of modern geography may be recapitulated as follows:

First, it is clear that geography overlaps with other disciplines. This is emphatically not a peculiarity or weakness of the subject, for the interdependence of disciplines is a main trend in the development of contemporary knowledge. Indeed, there is much to be said for the view that the organization of departments for the study of separate disciplines is largely a traditional convenience. They could probably be more usefully reorganized, or new departments added, certainly at the research level, on the basis of common fields of investigation. A great challenge to University education is to find ways and means of really breaking down the barriers of single disciplines. We need broader programmes of study and new avenues of effective interdisciplinary co-operation in teaching and research. These can be effected through the study of particular major culture realms (e.g. West European, Soviet, Chinese) or major fields of study such as urbanism or international relations. To any such research, geography is fundamental.

Second, in spite of its common ground with other sciences, as Carl Ritter pointed out one hundred years ago, geography must always keep clear its own goals. The last thirty years have seen a remarkable development of the objectives and methods of geographic research, and concepts need to be re-examined and reformulated. The history of the subject reveals that the essential basis of geographical work lies in the study of the areal association of phenomena on the earth, not in the exclusive associations of man-land relationships in the traditional sense of environmental relationships. *Geography is fundamentally the regional or chorological science of the surface of the earth.*

Third, this wide concept of geography as a chorological science, with its varied aspects on a footing of equality, is too wide for effective cultivation at the research level in all its aspects by one small group of scholars. If geography is limited to the earth as 'the home of man' or 'man on the earth' in terms of repetitive patterns of man-made relationships, then the physical environment must be evaluated in its relevance to man. Numerous facts and phenomena have an areal distribution which the professional geographer would regard outside his province. The field of geomorphology may be regarded as a distinct discipline, closely allied to geology. Climatology is closely allied to meteorology and there are now separate chairs in the latter. Oceanography, as a field of research, has long been voluntarily relinquished by geographers, though they naturally have a general interest in it. Plant geography, under the title of plant ecology, is assiduously pursued by botanists. Very few geographers have turned to questions of zoogeography, but the role of 'territory' of 'living-share' in the life of worms, birds and animals, has recently acquired great prominance. The history of the I.G.U. and current trends reveal concern with such highly specialized fields as limonology, palynology, geochronology, etc. On the human side, continental scholars, notably in Germany, have long turned their attention to the economic factors involved in the distribution of phenomena and to the geographical structure of economic groupings. *Wirtschaftskunde* has long been studied, elaborated and recently developed in the work of Lösch and, in the post-war years by scholars, primarily economists, in the United States (Isard, Garrison and Warntz) and in France (Ponsard). The geographical distribution of social groups developed by the human ecologists in the Chicago school of sociology of the twenties and their successors has produced a formidable body of factual and theoretical work. The political scientist following on the work of 'geopolitics' of the Swede, R. Kjellen, is also interested in the State as a geographical phenomenon, though less so than in the other two fields. We hear currently of human ecology, ekistics and regional science. The methods of quantitative analysis are also much in vogue. These are all 'systematic' or 'nomothetic' aspects of earth-associated phenomena, and are becoming to an increasing degree part of highly specialized fields of knowledge. What remains to the geographer as the distinct core of his expertise? Clearly the way in which these varied earth-bound phenomena are inter-related and react upon each other to form distinct areal associations of phenomena, as distinct environments, milieus, or spatial complexes. This is the essence of geographic work. It is the regional concept. The central concern of the regional concept is with problems of areal analysis, the objectives of which, as P. E. James, my esteemed American friend and colleague, has pointed out, may be descriptive, genetic or remedial. Descriptive studies assemble the data of an area in terms of their distribution and their areal coincidences, without necessarily pursuing their causes. Genetic studies are concerned with the processes involved in the development of spatial variables on the earth's surface, be they routes, grasslands or farm-steads. Remedial studies seek to detect the spatial changes and maladjustments of sociographic patterns, be they local government divisions or international frontiers. All these kinds of problems use the same basic approach of accurate descriptive analysis of areal groupings of phenomena, but each has differing objectives.

The term 'description' is currently frowned upon. As a reaction to old methods this is good, but it is unfortunate in the sense that descriptive studies of land, people and resources are urgently needed for wide areas of the earth alongside more specialized genetic studies of particular aspects. What we need is 'explanatory description' with clear objectives in the *co-variations of distributional data*. The term 'physical base', used in the training of geographers in Britain, is a hangover from the old environmental view. It should be dropped, unless its meaning be changed and recognition of the 'human base', as envisaged by Hans Bobek, be placed alongside, for geographic study seeks to determine and interpret the areal differences of lands and peoples on the earth's surface. Qualities of the earth should be evaluated in terms of their relevance to human use. Geographical holism means nothing as a geographical concept, unless it is tied down to rigorous yardsticks. If spatial studies of the earth's surface are so far removed as to be concerned with entirely different processes, then geography as a single discipline will be split down the middle or even into several sections. One single focus is needed for the professional geographer. This can only be the mode of the areal associations of phenomena on the earth in terms of localization and explanation as to origin, composition and spread.

Fourth, two points should be made with reference to Britain and America. There is need for greater concentration on the systematic fields, and for limited objectives in the study of the integration of phenomena in particular areas. This trend is very evident in both

French and German research. It is also abundantly evident that every professional geographer should specialize on one major world area and soak himself in the language and culture of the people at first hand. There is an appalling dearth of area specialists who have mastered the appropriate language, be it Russian, Arabic or Chinese, and who have lived and worked and roughed it for any length of time in some distant land. Work at home, and no less in foreign fields, has long been a tradition of German and French geography. Where are the specialists in Britain and America on the Soviet Union, Latin America, China, or even Western Europe?

On the other hand, it is surprising how little has been contributed in English to the theoretical development of the 'systematic' or 'topical' fields of geography[1]. This, too, can be explained by the lack of research in lands abroad, and a consequent lack of data, experience and perspective. This, however, has been a most significant trend in Germany and France in the post-war years. In Europe, regional monographs at home and in foreign fields have been paralleled by treatises on topical aspects, such as population, agrarian systems, industries, energy, religions, and circulation. It is also not surprising how unsatisfactory and elusive geographers in Britain and America have found the notion of 'regional synthesis' that has pursued juxtaposition, rather than the processes of spatial association. What is widely accepted as 'the regional concept' in the English-speaking world, is a generation out of date. The systematic fields must be developed and their expertise applied *in particular areas*. In other words, objectives must be limited and rigorously formulated in terms of geographical concepts. The findings of the systematic field must be put to work, in their repercussions on each other, within the framework of areal contexts in particular areas of the world.

The profession of geography faces both a pedagogical and a research challenge. George Chabot has written somewhere that geography is 'bursting out all over', stimulating and pursuing important new fields of enquiry in the natural and social sciences. Many geographers seem to find their work more in harmony with that of peripheral disciplines, to judge from the placement of their publications. All the more reason why geography should be more succinctly defined and pursued as to the content of its core. This demands action rather than the *malaise* that grows out of a defeatism for which there is, from a scholastic point of view, absolutely no justification. The remarkable growth of modern geography, as

[1] Fortunately, new publications in both Britain and America, reveal a change in this direction.

traced in this book, provides testimony to this judgement, but the purpose and concepts of geography, and the objectives of its practitioners, as teachers and researchers, need to be clearly and boldly articulated.

THREE NEW SIGNPOSTS

The field of geography in France today is presented in a large volume entitled *Géographie Générale* in the *Encyclopédie de la Pléiade*, published in 1966 and edited by André Journaux, Pierre Deffontaines, and Muriel Jean-Bruhnes Delamaare. Chapters are written by some thirty younger geographers (of the fourth generation!) who are among the most active of their contemporaries. The first part on Physical Geography seeks to present the natural processes that operate to create the *paysages naturels*. Depiction of these *paysages* of the world as a whole (on a generic basis) involves a synthesis of the climatic, pedological, biogeographic, and morphological elements. Thus, climatology and climatic zones, fluvial hydrology, mechanical and chemical erosion under different climatic conditions, the structural basis of land-forms, are the headings of the first chapters in the section on physical geography. This concludes with a portrayal of the *paysages naturels*. These are examined within the framework of the major climatic belts—temperate zone (an unfortunate choice of terms, without precise definition), glacial and periglacial zones, tropical and desert zones, littorals, and ocean floors. These 'natural regions', it is asserted, are the backcloth on which Man makes his imprint. A last chapter deals with the application of physical geography to problems of planning.

The second half of the book deals with human geography. Note well the expression used to express the objective—*la géographie de l'oeuvre paysageiste des hommes sur la terre*, that is, the impact of Man on the natural landscapes or regions. This part is organised in four sections, 'the human effectives' (*les effectifs humains*)—a term that was first adopted by Deffontaines in his monograph of 1932–, the development of natural resources, the geography of transport, and cultural geography.

Cultural geography contains chapters on the geography of literacy, of literature, the geography of leisure, the geography of religions, the geography of law, and political geography. The last, reported by Gottmann, is summarised on p. 248. The chapters on the human effectives include prehistory, population, pioneer fringes, cities, night and sleep (by Deffontaines), hunger, sanitation, and disease. The

281

chapters on the development and organisation of resources (*mise en valeur*) embrace collecting, hunting, and fishing communities; steppes and deserts of tropical lands; the American Arctic; the objectives of agriculture and pastoralism; the geography of forest, mining, industry, salt, and energy, and the geography of transport—land, sea, air, and telecommunications, This is a mere recital of chapter headings, but it certainly indicates the field and research prospects of geography today. The approach reveals shades of Bruhnes, Leplay, Vidal, and Sorre. It reflects the strong humanistic tradition in French geographic scholarship.

Certain geographers, as Pinchemel points out (p. 256), who are concerned about the rift in geography, seek to strengthen the mainstream as a single-focussed discipline with a scientific basis. This purpose is evidenced in two new and major treatises—Jean Labasse, *L'Organisation de l'Espace; Éléments de Géographie volontaire*, 1966, and Paul Claval, *Régions, Nations, Grands Espaces*, 1968. Labasse focusses his interpretation on the 'organisation of space' (*aménagement du territoire*,) or, alternatively, with reference particularly to the domestic front, 'environmental planning'. These are terms which the reader will recall have cropped up at many points on the preceding pages in the trends of current thought. Labasse seeks not only to clarify the scope and purpose of this approach, but also to apply it to the solution of practical problems of human organisation of space. Claval's work puts its emphasis on the quantitative analysis of the areal distribution and association of economic phenomena. This field of enquiry received its main impetus from German scholars in the thirties, notably A. Lösch and W. Christaller, whose major works are available in English (published in the United States). Their ideas strongly influenced the present writer in the thirties. American and British geographers, as well as planners, such as Constantin Doxiadis of Athens, Greece (who, in a *Profile* in the *New Yorker* several years ago admitted, at length, that his field of ekistics or the scientific study of settlement, was basically derived from the work of Christaller), and 'regional scientists', such as W. Isard in the United States, have awakened in the post-war years to the importance and opportunity for the application of such study to practical matters. But it needs to be put clearly on record that they all explicitly are building on the shoulders of the two German scholars who preceded them thirty odd years ago. Unfortunately, some assume, with evangelical zeal, that they are apostles of a 'new geography', which, they claim, is far removed from what is often condescendingly described as 'the traditional geography'. How far this attitude is justified the reader can judge for himself after reading this book. Labasse and Claval formulate the purpose and principles of areal economic analysis within the traditional framework, and apply them to a more effective interpretation of regional associations of phenomena over the earth—associations which Claval calls 'regions, nations, and major world areas'. Both these men use the techniques of quantitative areal analysis (and Labasse has done this admirably elsewhere in his monograph on the city-region of Lyons) to help push forward geography in the mainstream of its heritage, as I have tried to present it in these pages.

Subject Index

Subject Index

Cosmography, 4

Co-variation of distributional data, 279

Criteria of quantity, 64

Cultural and economic regions, 182–4

Cultural geography, 131, 139, 154, 162, 163, 169, 170, 176
See also Human geography

Cultural landscape (*Kulturlandschaft*), 70, 72, 128–35, 139, 172–3: and natural landscape, 159, 160; immobile and mobile forces, 129

Cultural spread or diffusion, 70

Culture areas, theory of, 64

Demography, 258, 259

Descriptive geography, 83–4, 116, 264, 279: explanatory, 279

Deutschland, conception of, 105–6, 108

Diffusion, 70: by migration or borrowing, 64

Distribution, 118: categories of, 86–7; co-variations of data, 279; general principles, 87; spatial, 87

Dualism in geography, 46, 51–3, 56

Dynamic approach, 84–6, 131

Earth features, forms of, 103–4

Earth science, *see Allgemeine Erdkunde* and Geophyiscs

Earth's surface, 12, 80–6, 116, 192, 213; areal differentiations or variations, 11, 12, 46, 116–17; entities, 180; features, 80–2; morphology of, *see* Geomorphology

Ecology and ecological approach, 46, 52, 154, 166, 168, 173, 236–8, 255, 278; regional ecology, 179–85; social ecology, 177; use of term ecological, 181–2

Economic geography, 109, 117, 118, 147, 150–1, 153–7, 171, 213–14, 219, 232, 247–9, 264, 278; economic and cultural regions, 182–4; economic landscape and 'formations', 154, 156

Ecosystems, 174, 182

Ecotopes, 181, 182, 183

Ecumene, the, 3, 4, 68

Ekistics, 278

Empirical (descriptive) approach, 11

Energy, geography of, 260

Environment, 235, 246: effects of physical, 68; geographical, 200–2, 210; natural, 131, 210

Erdkunde, 23–4, 46, 82, 116, 214: *physisiche*, 56, 98, 99

Erdoberfläche (earth's surface), 82–3, 85, 116

Erdräum, 83–4

Erdteile, 38–40, 44

Ethnogeography, 118

Ethnography, 19, 52, 62–4, 70, 85, 95, 139, 258

Eugenics, 63

Family life and organization, 199–203

'Fixed' forms, 39

Folk ideas, 63

Forest clearance and conservation, 13–14, 73–5, 127–8, 172–3

Functional approach, 173–4

General or universal geography, 6–9, 19–20, 31, 39, 45, 46, 51, 52, 56, 83–4, 94, 116, 169–75, 184: absolute, relative and comparative, 8
See also Allgemeine Erdkunde and *Allgemeine Geographie*

Genetic approach and force, 84–6, 119, 131, 139–40, 264, 279: monodynamic and polydynamic forms, 140

286

Subject Index

Subject Index

Subject Index

Subject Index

Transport geography, 147, 232

Units, geographical, 12, 15–18, 38, 40, 140–1: as spatial concept, 44; natural, 15–18, 38, 189; of landscape, 132–4, 140–1, physical land units, 191–2; structure and hierarchy, 163, 180–1

Unity: of nature and of man, 12, 30–1; organic unity of areas, 44; terrestrial or spatial, 38, 44

Universal geography(ies), 6–9, 19–20

Urbanism, geography of, 155, 171, 173, 235–6, 247–8, 259–61

Urlandschaft, 128

Viticulture, geography of, 244

Völkerkunde, 15, 17, 56, 58, 95, 139

Volksboden, 105, 108

Vulcanicity, 39, 68, 146, 147

Whole, organic and composite, 38

World areas, 207

World landscape, 128

World trade, geography of, 93

World units, hierarchy of, 163

Zoogeography, 278

Zoology and zoological geography, 117, 170

Zusammenhang, 42

Name Index

Name Index

Name Index

Name Index

Name Index

Geography of Europe (Gottmann), 248

Geography of Man, A (Hettner), 115

George, Pierre (b. 1909), 168, 194, 216, 241, 244–6: and advancement of status of geography, 245; and geography of energy, 260; and study of human societies, 246; and urbanism, 259; on natural and cultural environment, 246; popularizer of geography, 244–5

Gerland, Georg (1833–1919), 53, 54, 59, 94–6, 146: exclusive concern with physical earth, 95; radical views, 95

Germanisches Europa (Mendelssohn), 52

German Universities, development of geography in, 59–61

Gibert, A., 216

Gilbert, G. K., 120, 276

Girardin, P., 213

Glacken, Clarence, 12–14, 74

Gletscher der Vorzeit in den Carathen und in den Mittelgebirgen Deutschlands, Die (Partsch), 91

Gobineau, Arthur, 62

Goethe, 23, 30

Göttingen, University of, geographical tradition at, 55

Gottmann, Jean (b. 1915), 178–9, 194, 230–1, 233–4, 241, 246–9: and spatial organization of human groups, 248–9; and urbanization, 247–8, on economic and political geography, 247–9; on tradition vidalienne, 220–1; philosophy and procedure, 248–9

Götz, 118

Gourou, Pierre, 191, 195, 216, 221, 234, 240, 258

Gradmann, Robert (1865–1950), 60, 87, 100, 123, 135, 144–5,

159, 160, 164, 184: study of plant geography, prehistory and land settlement, 145, 172–3, 259; work on south Germany, 145

Grande Limogne auvergnate et bourbonnaise, La (Derruau), 261, 262

Grandes routes despeuples, Les (Demolins), 201–2

Granö, 178

Grant, Madison, 63, 203

Gravier, J. F., 260

Grenoble, Ecole de, 194, 236

Griechischen Landschaften, Die (Philippson), 143

Grisebach, August, 107

Grossen Fischerräume der Welt, Die (Bartz), 185

Grossen Herder Atlas, 166

Grundlagen der Landschaftkunde, Die (Passage), 121–2, 138, 140

Grundzüge der Allgemeinen Geographie (Philippson), 142

Grundzüge der Länderkunde (Hettner), 115, 123

Grundzüge der physischen Erdkunde (Supan), 99

Guide de l'Etudiant en Géographie (Cholley), 251

Guilcher, A., and oceanography, 257

Guthe, Hermann, 51, 52, 94

Gutsmuths, J. C. F., 36

Haack, Hermann, 161, 162

Haddon, A. C., 62, 224 n.

Haeckel, Ernst Heinrich, 65, 66

Hahn, Eduard, 64, 123, 152, 154

Hahn, F., 97

Halle, University of, geographical associations, 97

Handbuch der allgemeinen Erdkunde, unpublished MS by Ritter, 35

Handbuch der Geographischen Wissenschaft (ed. Klute), 115, 148, 172, 174, 176

296

Name Index

Hann, Julius, 68, 107
Hanotaux, Gabriel, 212, 243
Hardy, 191
Hartke, W., 60, 151, 157, 168, 170, 173, 267: and functional entities, 174
Hartshorne, R., 18, 43, 46, 95, 134–5, 175
Hassert, K. (1868–1947), 100, 123
Hassinger, Hugo (1877–1945), 104, 110, 147–8, 167, 172, 174, 176: and *Mitteleuropa*, 148; regional studies, 148
Haushofer, Karl, 71, 114, 162
Haute Durance et Ubaye (Péguy), 236, 257, 261
Hecataeus of Miletus, 4
Hedin, Sven, 79, 86
Hegel, 23
Hehn, Victor, 73
Heim, H, 68
Helmholtz, 30
Herbertson, A. J., 120, 138, 181, 197, 204, 206, 210, 226
Herbertson, Dorothy, 206
Herodotus, 4
Herzog, Th., 164, 165
Hettner, Alfred (1859–1941), xii, 20, 52, 53, 60, 61, 64, 82, 87, 89, 94–7, 100, 111–26, 130–5, 137, 139, 142, 147, 152–60, 163, 164, 173–5, 178, 185, 193, 226: conceptual approach, 117; controversy with Davis, 119–22, 192; life and works, 112–19; theory of geography, 115–19; travels, 112–13
Himly, A., 190, 194, 226
Hindernissen, Von den (Bucher), 18
Histoire de la Nation Française (ed. Hanotaux), 213, 243
Histoire naturelle, générale et particulière (Buffon), 12–13
Hohe Tatra zur Eiszeit (Partsch), 93
Holdrich, Thomas, 269

Homer, 3
Homme et la Terre, L' (Reclus), 222
Homme sur la Terre, La (Sorre), 237
Hommes et leurs travaux dans les pays de la Moyenne Garonne, Les (Deffontaines), 216–18, 242
Hommeyer, H. G., 16
Hudson, 181
Humboldt, Alexander von (1769–1859), xi, 18, 21–34, 38, 42, 43, 52, 54, 56–8, 66, 72, 73, 82, 84, 102, 113, 147, 166, 184, 189: and areal associations, 24; and climatology, 24; and landscape, 26; and plant geography, 25; contribution to geography, 31; divisions of knowledge, 23–4; fieldwork, 26–7; regionalism, 28; scientific procedure, 23; travels, 22; work on Mexico, 26–8
Huntingdon, Ellsworth, 73

Industrie des Alpes Françaises, L' (Veyret-Vernier), 236
International Cartography Association, 275
International Geographical Union, 230, 267, 271–7: and human geography, 276–7; and physical geography, 276; bibliographical work, 275, 276; change in scope and depth of its field, 275–6, congresses, 271–7; hundred years of its history, 276–7; national committees, 272
International Map, 271, 272
Introductio in Universam Geographiam (Cluverius), 9–10
Irrigation (Brunhes), 212
Isard, 278

Jaeger, Fritz, 113, 124, 155
James, Preston E., 131, 133, 179, 181, 182, 233, 279

297

Name Index

Joanne, Adolphe, 225
Joerg, W. L. G., 208
Johnson, D. W., 231, 239, 276
Jones, Wellington, 133
Joule, 30
Journaux, André, 243, 246, 248, 261: on post-war trends in France, 221
Juillard, Et., 194, 259, 261: and agrarian landscape, 258

Kant, Immanuel (1724–1804), 3, 10–11, 95: division of geography, 11
Kapp, Ernst, 51
Kaubler, R., 127
Keane, A. H., 62, 222, 224 n.
Keltie, J. Scott, 270
Kiepert, Heinrich (1818–99), 51, 53, 54, 79, 89–93
Kirchhoff, Alfred (1838–1907), 59, 67, 87, 96–8, 107, 112, 126; and *Länderkunde*, 97, 185; and methodology, 97; and training of teachers, 97
Kirsten, E., 143
Kjellen, R., 71, 278
Kleine Landeskunde der Provinz Schlesien, Eine (Partsch), 91
Kleine Schriften (Ratzel), 67
Klute, Fritz, 94, 115, 148, 162
Kohl, J. G., 51
Kolb, Albert, 113, 160, 185
Kraus, Th., 174, 176
Krebs, Norbert (1876–1947), 60, 87, 93, 101, 104–6, 108, 109, 130, 141, 148, 152–3, 163, 167, 172: and *Lebensraum*, 153; regional studies, 152, 153
Krümmel, Otto, 68, 94
Kropotkin, Petr, 73
Kuske, Bruno, 154

Labasse, Jean, 194, 259
Länderkunde der Österreichischen

Alpe (Krebs), 87
Länderkunde der südeuropäischen Halbinseln (Fischer), 96
Länderkunde von Europa (Kirchhoff's series), 96–8, 107
Landes-und Volkskunde der Provinz Schlesiens (Partsch), 91
Landry, A., 259
Landschaftsgürtel der Erde, Die (Passarge), 138
Landschaft und Wirtschaft in Schweden (Credner), 156
Landwirtschaftsgeographische Arbeitsgemeinschaft, 157
La Plata Länder (Wilhelmy and Rohmeder), 185
Lauer, W., 158, 159
Lautensach, Hermann (b. 1886), 105, 131, 132, 150, 160–4, 169, 185: and cultural geography, 163; and physical geography, 163; and regional concept, 176–9; hierarchy of world units, research in regional geography, 161, 163; studies of Iberian Peninsula and of Korea, 162, 163, 177–9; theory and practice of geography, 163
Lebensräume in Kampf der Kulturen (Schmitthenner), 177
Lebensraumfragen europäischer Völker, 158
Leçons de géographie physique (De Lapparent), 192
Lefebvre, Th., 216
Lehmann, Herbert, 60, 104–5, 142, 143
Lehmann, O., 93
Lehrbuch (Supan), 98
Lehrbuch der Alten Geographie (Kiepert), 90
Lehrbuch der Geographie (Guthe and Wagner), 94
Leibnitz, 12
Leipoldt, Gustav, 56
Leipzig University, geographical tradition, 79

298

Name Index

Name Index

Name Index

301

Name Index

302

Name Index

303

Name Index

Tiessen, Ernst, 78
Tocqueville, 239
Tomaschek, Wilhelm (1841–1901),
110, 148
Traité de Géographie Physique
(de Martonne), 230
Traité de Géographie Urbaine
(Chabot and Beaujeu-Garnier),
259
Travels in the Columbian Andes
(Hettner), 112
Travels of a Naturalist (Ratzel), 66
Tricart, J. 251, 255, 261
Troll, Carl (b. 1899), 60, 104, 109,
115, 131, 143, 147, 151, 157,
161–6, 169: and air photography,
166; and climatology, 166;
and glacial morphology, 164–6;
and methodology of geography,
166; and study of small areas,
183; expeditions to Africa and
Himalaya, 165–6; founds
Erdkunde, 166; on association of
cultural and economic regions,
182–4; on hierarchy of units,
180–2; on regional concept,
179–80; president of International Geographical Union,
274; researches in Andes, 165
Tropen, Die (Sapper), 147
Tylor, Sir E. B., 64

Uhlig, Carl, 113
Universal Geography (Malte-
Brun), 19
Unstead, J. F., 181
Unwin, Raymond, 204

Vacher, A., 215
Vacher de Lapouge, Georges,
62–3
Val de Loire, Le (Dion), 216, 218,
243, 244
Vallaux, Camille, 176, 194, 213,
215
Van Vuuren, 176
Varenius, Bernhard (1622–50),

3, 6–10: definition and division
of geography, 8
*Vergleichende Allgemeine
Erdkunde* (Kapp), 51
Vergleichende Länderkunde
(Hettner), 115
Vergleichende Länderkunde
(Krebs and Lautersach), 63
Vergleichende Landschaftskunde
(Passarge), 138, 140
*Vergletscherung der Deutschen
Alpen, Die* (Penck), 91, 102
Veyret-Vernier, G., 236, 259, 261
Vidal de la Blache, Paul (1845–
1918), xii, 20, 53, 64, 72, 123,
124, 126, 139, 190–5, 205–12,
224–6, 228–31, 234–9, 241, 243,
246, 250, 251, 253, 254, 263,
265, 270: and *genres de vie*, 201,
207, 208, 210–11, 233, 236,
238; and geographical possibilism, 193, 203, 239; and
geography as distinct discipline,
210; and natural or geographic
environment, 210; and
physiognomic des paysages, 210;
and production of regional
monographs, 210, 214–15;
concept of human geography,
193; initiates *Geographie
Universelle*, 209–10, 218–19;
tradition vidalienne, 193, 208,
215, 219–21
Vie forestière en Slovaquie, La
(Deffontaines), 242
Vienna, University of, sequence
of appointments, 109–10
*Vie Pastorale dans les Alpes
Françaises, La* (Arbos), 236
Virginia at mid-century
(Gottmann), 179, 247
Vogel, Walter, 109
Völkerkunde (Peschel), 58, 62
*Völkerkunde (The History of
Mankind)* (Ratzel), 62, 64, 66
Volz, Wilhelm (1870–1958), 60,
100

304

Name Index